T0371991

Systems and Implemented Technologies for Data-Driven Innovation, Addressing Data Spaces and Marketplaces Semantic Interoperability Needs

i3-MARKET Series - Part II: Data Economy, Models, Technologies and Solutions

RIVER PUBLISHERS SERIES IN COMPUTING AND INFORMATION SCIENCE AND TECHNOLOGY

Series Editors:

The "River Publishers Series in Computing and Information Science and Technology" covers research which ushers the 21st Century into an Internet and multimedia era. Networking suggests transportation of such multimedia contents among nodes in communication and/or computer networks, to facilitate the ultimate Internet.

Theory, technologies, protocols and standards, applications/services, practice and implementation of wired/wireless networking are all within the scope of this series. Based on network and communication science, we further extend the scope for 21st Century life through the knowledge in machine learning, embedded systems, cognitive science, pattern recognition, quantum/biological/molecular computation and information processing, user behaviors and interface, and applications across healthcare and society.

Books published in the series include research monographs, edited volumes, handbooks and textbooks. The books provide professionals, researchers, educators, and advanced students in the field with an invaluable insight into the latest research and developments.

Topics included in the series are as follows:-

- Artificial intelligence
- Cognitive Science and Brian Science
- Communication/Computer Networking Technologies and Applications
- Computation and Information Processing
- Computer Architectures
- Computer networks
- Computer Science
- Embedded Systems
- Evolutionary computation
- Information Modelling
- Information Theory
- Machine Intelligence
- Neural computing and machine learning
- Parallel and Distributed Systems
- Programming Languages
- Reconfigurable Computing
- Research Informatics
- Soft computing techniques
- Software Development
- Software Engineering
- Software Maintenance

For a list of other books in this series, visit www.riverpublishers.com

Systems and Implemented Technologies for Data-Driven Innovation, Addressing Data Spaces and Marketplaces Semantic Interoperability Needs

i3-MARKET Series - Part II: Data Economy, Models, Technologies and Solutions

Editors

Martín Serrano
Achille Zappa
Waheed Ashraf
Pedro Maló
Márcio Mateus
Edgar Fries
Iván Martínez
Alessandro Amicone
Justina Bieliauskaite
Marina Cugurra

LONDON AND NEW YORK

Published 2024 by River Publishers
River Publishers
Alsbjergvej 10, 9260 Gistrup, Denmark
www.riverpublishers.com

Distributed exclusively by Routledge
4 Park Square, Milton Park, Abingdon, Oxon OX14 4RN
605 Third Avenue, New York, NY 10017, USA

Systems and Implemented Technologies for Data-Driven Innovation, Addressing Data Spaces and Marketplaces Semantic Interoperability Needs / by Martín Serrano, Achille Zappa, Waheed Ashraf, Pedro Maló, Márcio Mateus, Edgar Fries, Iván Martínez, Alessandro Amicone, Justina Bieliauskaite, Marina Cugurra.

ISBN: 978-87-7004-171-3 (hardback)
978-10-4009-096-1 (online)
978-10-0349-835-3 (master ebook)

DOI: 10.1201/9788770041713

Contents

Preface

Data is the oil in today's global economy. The vision of the i3-MARKET book series is that the fast-growing data marketplaces sector will mature, with a large number of data-driven opportunities for commercialization and activating new innovation channels for the data.

A new data-as-a-service paradigm where the data can be traded and commercialized securely and transparently and with total liberty at the local and global scale directly from the data producer is necessary. This new paradigm is the result of an evolution process where data producers are more active owners of the collected data while at the same time catapulting disruptive data-centric applications and services. i3-MARKET takes a step forward and provides support tools for this maturity vision/process.

i3-MARKET is a fully open source backplane platform that can be used as a set of support tools or a standalone platform implementation of data economy support services. i3-MARKET is the result of shared perspectives from a representative global group of experts, providing a common vision in data economy and identifying impacts and business opportunities in the different areas where data is produced.

Data economy is commonly referring to the diversity in the use of data to provide social benefits and have a direct impact in people's life. From a technological point of view, data economy implies technological services to underpin the delivery of data applications that bring value and address the diverse demands on selling, buying, and trading data assets. The demand and the supply side in the data is increasing exponentially and it is being demonstrated that the value that the data has today is as relevant as any other tangible and intangible assets in the global economy.

This publication is supported with EU research funds under grant agreement i3-MARKET-871754. Intelligent, Interoperable, Integrative and deployable open source MARKETplace with trusted and secure software tools for incentivising the industry data economy and the Science Foundation Ireland research funds under grant agreement SFI/12/RC/2289_P2. Insight SFI Research Centre for Data Analytics. The European Commission and the SFI support for the production of this publication does not constitute an endorsement of the contents, which reflect the views only of the authors, and the Commission, the SFI or its authors cannot be held responsible for any use which may be made of the information contained therein.

Dr. J. Martin Serrano O.
i3-MARKET Scientific Manager and Data Scientist
Adjunct Lecturer and Senior SFI Research Fellow at University of Galway
Data Science Institute - Insight SFI Research Centre for Data Analytics
Unit Head of Internet of Things, Stream Processing and Intelligent Systems Research Group
University of Galway, www.universityofgalway.ie | Ollscoil na Gaillimh
<jamiemartin.serranoorozco@universityofgalway.ie>
<martin.serrano@insight-centre.org>
<martin.serrano@nuigalway.ie>

Who Should Read this Book?

General Public and Students

This Book is a unique opportunity for understanding the future of data spaces and marketplace assets, their services, and their ability to identify different methodologies indicators and the data-driven economy from a human-centric perspective supports the digital transformation.

Entrepreneurs and SMEs

This Book is a unique opportunity for understanding the most updated software tools to innovate, increase opportunities, and increase the power of innovation into small and entrepreneurs to meet its full potential promoting participation across the data economy values and evolution of society towards a single digital strategy.

Technical Experts and Software Developers

This book is a guide for technolgy experts and open source enthusiast that includes the most recent experiences in Europe towards innovating software technology for for the financial and banking sectors.

Data Spaces & Data Markeplaces Policy Makers

This Book represent a unique offering for non-technical experts but that participates in the data economy process and the core data economy servicesto enable the sharing of innovation and new services across data spaces and marketplaces such as policy makers and standardisation organisatiosna and groups

What is Addressed in the i3-MARKET Book Series?

"Concepts and Design Innovations for the Digital Transformation of Data Spaces and Data Marketplaces"

In the first part of the i3-MARKET book series, we begin by discussing the principles of the modern data economy that lead to making the society more aware about the value of the data that is produced everyday by themselves but also in a collective manner, i.e., in an industrial manufacturing plant, a smart city full of sensors generating data about the behaviours of the city and their inhabitants and/or the wellbeing and healthcare levels of a region or specific locations, etc. Data business is one of the most disruptive areas in today's global economy, particularly with the value that large corporates have embedded in their solutions and products as a result of the use of data from every individual.

"Systems and Implemented Technologies for Data-driven Innovation, Addressing Data Spaces and Marketplaces Semantic Interoperability Needs"

In the second i3-MARKET series book, we start reviewing the basic technological principles and software best practices and standards for implementing and deploying data spaces and data marketplaces. The book provides a definition for data-driven society as: *The process to transform data production into data economy for the people using the emerging technologies and scientific advances in data science to underpin the delivery of data economic models and services.* This book further discusses why data spaces and data marketplaces are the focus in today's data-driven society as the trend to rapidly transforming the data perception in every aspect of our activities. In this book, technology assets that are designed and implemented following the i3-MARKET backplane reference implementation (WebRI) that uses open data, big data, IoT, and AI design principles are introduced. Moreover, the series of software assets grouped as sub-systems and composed by software artefacts are included and explained in full. Further, we describe i3-MARKET backplane tools and how these can be used for supporting marketplaces and its components including details of available data assets. Next, we provide a description of solutions developed in i3-MARKET as an overview of the potential for being the reference open source solution to improve data economy across different data marketplaces.

"Technical Innovation, Solving the Data Spaces and Marketplaces Interoperability Problems for the Global Data-driven Economy"

In the third i3-MARKET series book, we are focusing on including the best practices and simplest software methods and mechanisms that allow the i3-MARKET backplane reference implementation to be instantiated, tested, and validated even before the technical experts and developers community decide to integrate the i3-MARKET as a reference implementation or adopted open source software tools. In this book, the purpose of offering a guide book for technical experts and developers is addressed. This book addresses the so-called industrial deployment or pilots that need to have a clear understanding of the technological components and also the software infrastructures, thus it is important to provide the easy-to-follow steps to avoid overwhelm the deployment process.

i3-MARKET has three industrial pilots defined in terms of data resources used to deploy data-driven applications that use most of the i3-MARKET backplane services and functionalities. The different software technologies developed, including the use of open source frameworks, within the context of the i3-MARKET is considered as a bill of software artefacts of the resources needed to perform demonstrators, proof of concepts, and prototype solutions. The i3-MARKET handbook provided can actually be used as input for configurators and developers to set up and pre-test testbeds, and, therefore, it is extremely valuable to organizations to be used properly.

What is Covered in this i3-MARKET Part II Book?
"Systems and Implemented Technologies for Data-Driven Innovation, Addressing Data Spaces and Marketplaces Semantic Interoperability Needs"

Data Economy is commonly referring to the diversity in The use of data to provide social benefits and have a direct impact in people's life, from a technological point of view data economy implies technological services to underpin the delivery of data applications that bring value and addressed the diverse demands on selling, buying and trading data assets. The demand and the supply side in the data is increasing exponentially and it is being demonstrated that the value that the data has today is as relevant as any other tangible and intangible assets in the global economy. In this second book it is further discuss why Data is the focus in current technological developments towards digital markets and the meaning for data being the next asset to appear evolution in trading markets and at the same time it focuses on introducing the i3-MARKET technology and the proposed solutions.

This book further discusses why data spaces and data marketplaces are the focus in today's data-driven society as the trend to rapidly transforming the data perception in every aspect of our activities. In this book, technology assets that are designed and implemented following the i3-MARKET Backplane reference architecture (RA) that uses open data, big data, IoT, and AI design principles are introduced. Moreover, the series of software assets grouped as sub-systems and composed by software artefacts are included and explained in full.

Furthermore, this book series describes i3-MARKET Backplane tools and how these can be used for supporting marketplaces and its components including details of available data assets. Next, we provide a description of solutions developed in i3-MARKET as an overview of the potential for being the reference open-source solution to improve data economy across different data marketplaces.

Acknowledgements

Immense thanks to our families for their incomparable affection, jollity, and constant understanding that scientific career is not a work but a lifestyle, for encouraging us to be creative, for their enormous patience during the time away from them, invested in our scientific endeavours and responsibilities, and for their understanding about our deep love to our professional life and its consequences – we love you!

To all our friends and relatives for their comprehension when we had no time to spend with them and when we were not able to join in time because we were in a conference or attending yet another meeting and for their attention and the interest they have been showing all this time to keep our friendship alive; be sure, our sacrifices are well rewarded.

To all our colleagues, staff members, and students at our respective institutions, organizations, and companies for patiently listening with apparent attention to the descriptions and progress of our work and for the great experiences and the great time spent while working together with us and the contributions provided to culminate this book series project. In particular, thanks to the support and confidence from all people who believed this series of books would be finished in time and also to those that did not trust on it, because, thanks to them, we were more motivated to culminate the project.

To the scientific community who is our family when we are away and working far from our loved ones, for their incomparable affection, loyalty, and constant encouragement to be creative, and for their enormous patience during the time invested in understanding, presenting, and providing feedback to new concepts and ideas – sincerely to you all, thanks a million!

Martín Serrano on Behalf of All Authors

List of Figures

List of Tables

List of Contributors

Achille, Zappa, *NUIG, Ireland*

Alessandro, Amicone, *GFT, Italy*

Andrei, Coman, *Siemens SRL, Romania*

Andres, Ojamaa, *Guardtime, Estonia*

Angel, Cataron, *Siemens SRL, Romania*

Antonio,Jara, *Libellium/HOPU, Spain*

Birthe, Boehm, *Siemens AG (Erlangen), Germany*

Borja, Ruiz, *Atos, Spain*

Bruno, Almeida, *UNPARALLEL, Portugal*

Bruno, Michel, *IBM, Switzerland*

Carlos Miguel, Pina Vaz Gomes, *IBM, Switzerland*

Carmen, Pereira, *Atos, Spain*

Chi, Hung Le, *NUIG, Ireland*

Deborah, Goll *Digital SME, Belgium*

Dimitris, Drakoulis, *Telesto, Greece*

Edgar, Fries, *Siemens AG (Erlangen), Germany*

Fernando, Román García, *UPC, Spain*

Filia, Filippou, *Telesto, Greece*

George, Benos, *Telesto, Greece*

German, Molina, *Libellium/HOPU, Spain*

Hoan, Quoc, *NUIG, Ireland*

Iosif, Furtuna, *Siemens SRL, Romania*

Isabelle, Landreau, *IDEMIA, France*

Ivan, Martinez, *Atos, Spain*

James, Philpot, *Digital SME, Belgium*

Jean Loup, Depinay, *IDEMIA, France*

Joao, Oliveira, *UNPARALLEL, Portugal*

Jose, Luis Muñoz Tapia, *UPC, Spain*

Juan Eleazar, Escudero, *Libellium/HOPU, Spain*

Juan, Hernández Serrano, *UPC, Spain*

Juan , Salmerón, *UPC, Spain*

Justina, Bieliauskaite *Digital SME, Belgium*

Kaarel, Hanson, *Guardtime, Estonia*

Lauren, Del Giudice, *IDEMIA, France*

Luca, Marangoni, *GFT, Italy*

Lucas, Asmelash, *Digital SME, Belgium*

Lukas, Zimmerli, *IBM, Switzerland*

Márcio, Mateus, *UNPARALLEL, Portugal*

Marc, Catrisse, *UPC, Spain*

Mari, Paz Linares, *UPC, Spain*

Maria, Angeles Sanguino Gonzalez, *Atos, Spain*

Maria, Smyth, *NUIG, Ireland*

Marina, Cugurra, *ETA Consulting*

Marquart, Franz, *Siemens AG (Erlangen), Germany*

Martin, Serrano, *NUIG, Ireland*

Mirza, Fardeen Baig, *NUIG, Ireland*

Oxana, Matruglio, *Siemens AG (Erlangen), Germany*

Pascal, Duville, *IDEMIA, France*

Pedro, Ferreira, *UNPARALLEL, Portugal*

Pedro, Malo, *UNPARALLEL, Portugal*

Philippe, Hercelin, *IDEMIA, France*

Qaiser, Mehmood, *NUIG, Ireland*

Rafael, Genés, *UPC, Spain*

Raul, Santos, *Atos, Spain*

Rishabh, Chandaliya, *NUIG, Ireland*

Rupert, Gobber, *GFT, Italy*

Stefanie, Wolf, *Siemens AG (Erlangen), Germany*

Stratos, Baloutsos,*AUEB,Greece*

Susanne, Stahnke, *Siemens AG (Erlangen), Germany*

Tanel, Ojalill, *Guardtime, Estonia*

Timoleon, Farmakis, *AUEB, Greece*

Tomas, Pariente Lobo, *Atos, Spain*

Toufik, Ailane, *Siemens AG (Erlangen), Germany*

Víctor, Diví, *UPC, Spain*

Vasiliki, Koniakou, *AUEB, Greece*

Yvonne, Kovacs, *Siemens SRL, Romania*

List of Abbreviations

API	Application program interface
AUEB	Athens university of economic and business
CRS	Conflict-resolver service
DAML	Digital asset modelling language
Dapp	Decentralized application
DC	Data consumer
DCAT-AP	DCAT application profile
DID	Distributed identifier
DLT	Distributed ledger technology
DP	Data provider
DSA	Data sharing agreement
EVM	Ethereum virtual machine
GDPR	General data protection regulation
HW	Hardware wallet
IBAN	International bank account number
IRI	Internationalised resource identifier
JSON	JavaScript object notation
JWA	JSON web algorithms
JWK	JSON web key
JWS	JSON web signature
NRP	Non-repudiation protocol
ODIC	OpenID connect
OWL	Ontology web language
PoA	Proof of authority
PoO	Proof of origin
PoP	Proof of publication

PoR	Proof of reception
PoW	Proof of work
QoS	Quality of service
RA	Reference architecture
RDF	Resource description framework
RP	Relying party
RPC	Remote procedure call
SCM	Smart contract manager
SDK	Software development kit
SEA	Service execution agreement
SEED	Semantic engine
SLA	Service-level agreement
SPARQL	SPARQL protocol and RDF query language
SSI	Self-sovereign identity
SW	Software wallet
TM	Translation memory
TPMs	Trusted platform modules
TRL	Technology readiness level
TRN	Transaction reference number
TSP	Trust, security, and privacy
TTPA	Trusted third-party auditor
Turtle	Terse RDF triple language
UI	User interface
URI	Uniform resource identifier
VC	Verifiable credentials
W3C	World wide web consortium

1

Reference Architecture

The overall picture is the description of the system including the main building block or subsystems. Figure 1.1 shows the overall component diagram for i3-MARKET.

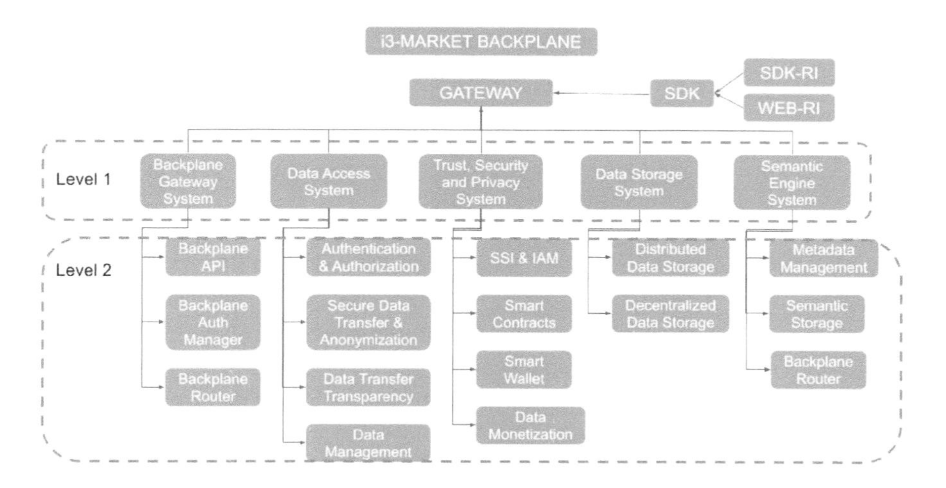

Figure 1.1 Overall system.

1.1 Level 1

Backplane gateway system:
The Backplane Gateway system is the building block in charge of offering to all participants and marketplaces access to the Backplane system. The goal of the Backplane API is therefore twofold: on one hand, it serves an integrated API endpoint for all the i3-MARKET services offered by i3-MARKET and

implemented in the respective building blocks. On the other hand, it provides secure mechanisms for preventing not allowed accesses.

In terms of public interfaces, the functionality integrated by the Backplane Gateway is exposed throughout the Backplane API.

In terms of internal connections with other i3-MARKET building blocks, the Backplane Gateway system has secure communication with the rest of the subsystems to integrate their services into the Backplane API. For this integration, any service must have a complete specification following the OpenAPI Specification 3.0.

Data access system:
The data access system is the building block in charge of allowing data consumers obtain access to the data offered by the data providers.

It exposes its functionality, publicly, through the secure data access API.

This data access system is securely linked with the Backplane API in order to guarantee two main issues.

Ensuring all the involved stakeholders have signed the required contracts and monitoring the quantities of exchanged data assets for the token-based monetization service.

Trust, security, and privacy system:
The trust, security, and privacy (TSP) is the building block in charge of providing the self-sovereign identity, access management, contracting, consents, accounting, and payments blockchain-based solutions managed in the i3-MARKET system in order to guarantee the desired levels of trust, security, and privacy for federated data markets.

The TSP system exposes its functionally, publicly, through the Backplane API.

In terms of dependencies with other existing building blocks, the TSP interacts with the decentralized ledger of the data storage system and with the data access system for the monetization of the data assets.

Semantic engine system:
The semantic engine system is the building block in charge of providing the needed semantic data models for making possible the consumers and applications understand the meaning of the data exchanged between different stakeholders. Apart from that, the semantic engine will allow the participants to take advantage of this semantic data model by means of providing a metadata management in charge of registering, offering, and performing queries for discovery purposes.

All this metadata management and query functionality is exposed, publicly, through the Backplane API.

In terms of dependencies with other building blocks, the semantic engine mainly interacts with the storage system for storing the offering descriptions.

Data storage system:
The data storage system is the building block in charge of storing common data shared across all participant instances.

It interacts with mostly all main building blocks, especially with the semantic engine system for performing the synchronization between semantic repositories and distributed storage and with the trust, security, and privacy for instantiating and executing smart contracts in the blockchain-based decentralized storage.

1.2 Level 2

Backplane gateway system – general description:
The Backplane Gateway has two main purposes:

Single entrypoint:

The Backplane offers a single set of endpoints, allowing clients to interact with all the services offered by the i3-MARKET project through a single API. This allows the whole system to have a modular and distributed architecture, while providing the ease of use of a single common interface.

Auth management:
All authentication and authorization flows are centralized and managed by the Backplane Gateway so that the subsystems are protected without them needing to handle their own auth flows. The actual authentication/authorization is delegated to specialized subsystems. This centralization also allows the support of several auth flows (and the addition of new ones) transparent to the several subsystems.

Besides these main purposes, the presence of the Backplane Gateway provides several benefits.

It allows the subsystems to be isolated and not exposed publicly so that they can only be accessed by the Backplane or other subsystems.External connections are all handled by the Backplane, making connection security and encryption simpler and more straightforward.The nature of the Backplane functionalities makes it easily scalable and replicable, making the addition of new Backplane instances transparent for both the subsystems and the clients. The Backplane Gateway is shown in Figure 1.2.

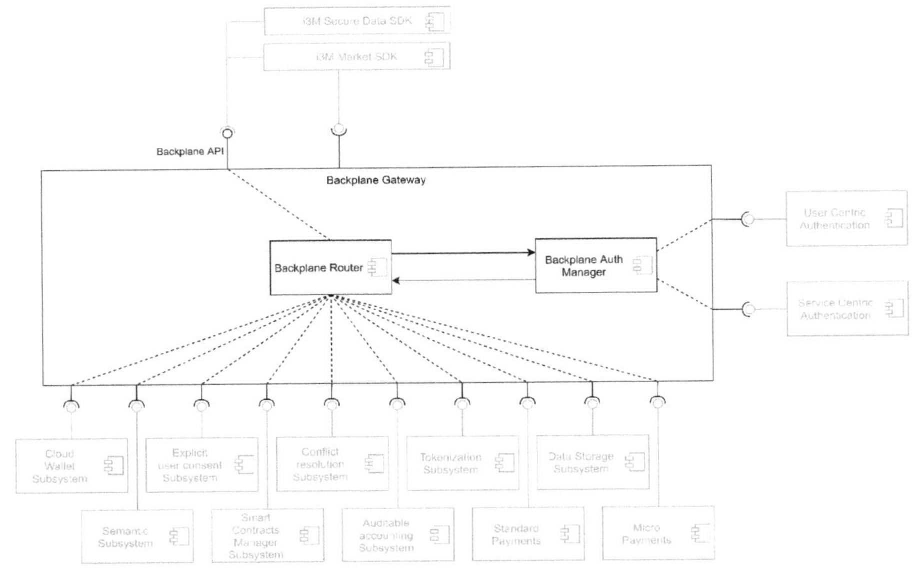

Figure 1.2 The Backplane Gateway component diagram.

Inner building blocks:
The Backplane Gateway consists of three main components.

Backplane API:
The Backplane API is the set of endpoints exposed by the Gateway. It comprises all the publicly available endpoints of the subsystems integrated with the Backplane as well as a few other endpoints, belonging to the Backplane itself, used in the authentication/authorization flows.

The API follows the OpenAPI Specification 3.0, and the endpoints corresponding to each subsystem are generated automatically based on the subsystem's own OpenAPI specification.

Backplane auth manager:
The Backplane auth manager is responsible for handling the authentication and the authorization required for the endpoints of the different subsystems. The actual auth processes are delegated to the corresponding subsystem (user/service-centric authentication subsystem), depending on the requirements of the endpoint and client.

Backplane router:
The Backplane router is the component of the Backplane responsible for the forwarding of the incoming requests to the several subsystems. It also checks

whether the endpoint requires authentication/authorization, invoking the auth manager if it does.

Data access system – general description:
The data access system consists of four main subsystems for implementing the transfer of data. Figure 1.3 shows the subsystems.

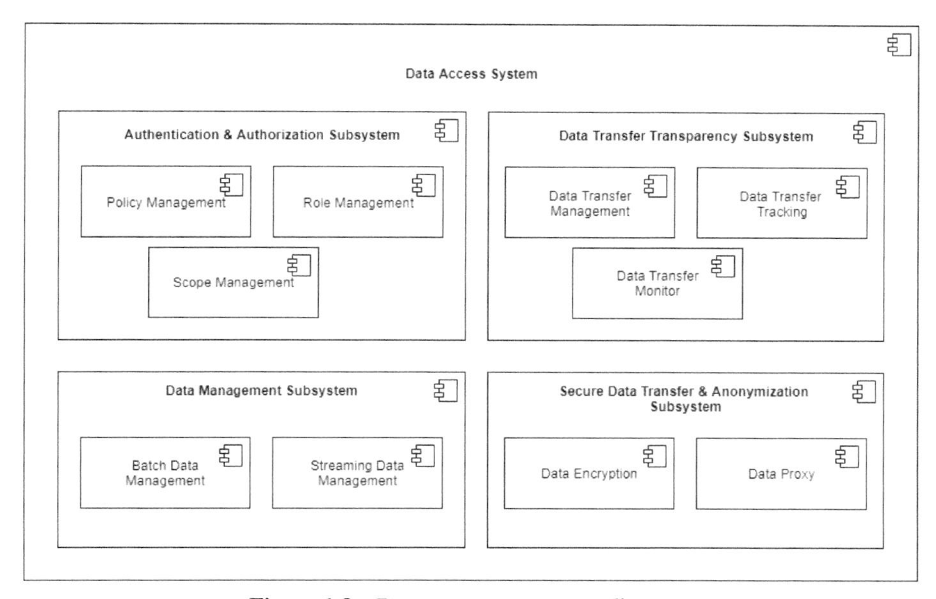

Figure 1.3 Data access component diagram.

Inner building blocks:

Authentication and authorization:
The data access system will assist i3-MARKET with the following capabilities for the exchanged data assets:

- Authentication: Verifies the identity of the user against the i3-MARKET Backplane. 7
- Authorization: Verifies the permissions the authenticated user has in the i3-MARKET platform allowing to perform authorized actions and grants access to resources.

Authentication and authorization subsystem has the following scope management:

- Policy management: Policy is a set of rules that defines how to protect the assets in order to provide trust, security, and privacy. The policy management component is in charge of enforcing the rule set provided by i3-MARKET Backplane inside of the data access system. The responsibilities of the policy management module are:

  - intercept access attempts;
  - check attempt against rule set;
  - grant access to permitted assets.

- Role management: A role is a set of policies attached to an entity in order to define the access that entity has within the i3-MARKET data access system. The role management component is in charge of fetching the list of policies and verifying them against the data access system. The responsibilities of the role management module are:
  - get the list of policies associated with role from Backplane;
  - verify role access by invoking policy management;
  - allow or deny functionalities.

Secure data transfer and anonymization:
The secure data transfer and anonymization subsystem has the following components:

- Data encryption: The responsibilities of the data encryption module are:
  - key generation and exchange;
  - transfer data in an encrypted way between endpoints;
  - decrypt data on the consumer side.
- Proxy: The proxy needs to be used when the identity of the data provider needs to be hidden. This feature is optional; there is no need to implement it if there is no specific requirement referring to the anonymity of the data provider. The responsibilities of the proxy module are:
  - activate the proxy;
  - configure the parameters to hide the identity;
  - data transfer goes through the proxy.

Data transfer transparency:
The data transfer transparency subsystem has the following components:

- Data transfer management: This component is responsible for the management of the connection between the provider and the consumer and implements the following functionalities:

  - initialize the connection;
  - resume the connection;
  - finalize the connection.
- Data transfer tracking: This component implements the following operation:
  - measure the amount of transferred data.
- Data transfer monitor: The information about how much data was transferred, when the data transfer was initiated, and when it was completed is monitored, and the following operations are triggered:
  - inform the i3-MARKET Backplane that the data transfer was performed and reports how much data was transferred;
  - invoke the linked smart contract.

Data management:
Two methods for data transfer are supported by data access API, which are supported by the following modules:

- Batch data transfer management: One-time data transfer for one chunk of data in a session with the following methods:
  - request data;
  - transfer data.
- Data stream management: Continuous transfer of data based on a subscription, e.g., publish/subscribe mechanism:
  - subscribe to an offering;
  - trigger data transfer – on the producer side;
  - get data – on the consumer side;
  - unsubscribe.

Trust, security, and privacy system – general description:
One of the pillars of the i3-MARKET Backplane is the "trust, security, and privacy" system, which leverages the blockchain technologies to ensure trust, security, and privacy by design, providing the following building blocks as shown in Figure 1.4:

- an identity and access management system based on decentralized/self-sovereign identity and Verifiable Credentials;
- smart wallets with different levels of security (Cloud/HW Wallet);

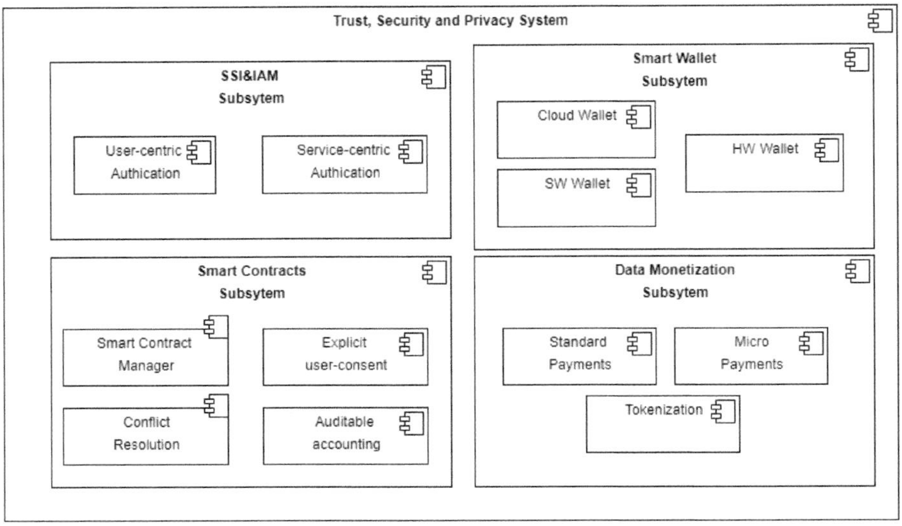

Figure 1.4 Trust, security, and privacy component diagram.

- smart contracts to record, operate, and manage in a trusted and transparent way the agreements between the different stakeholders and particularly the explicit user consent of the data owner;
- a data monetization system based on crypto currency for secure, trusted, and cost-effective peer-to-peer payments.

Inner building blocks:

SSI & IAM:

The SSI&IAM building block provides an authentication and authorization mechanism to access i3-MARKET Backplane and stakeholder resources.

The user-centric authentication component implements the self-sovereign identity paradigm ensuring that:

- identity and personal data are stored with the user;
- claims and attestations can be issued and verified between users and trusted parties;
- users selectively grant access to data, and data only needs to be verified a single time.

The service-centric authentication component makes the data marketplaces of the network able to provide their users a distributed identity they own and can use with other stakeholders.

Both the authentication mechanisms follow the OpenID Connect standard to allow wide commercial acceptance.

Smart contracts:
The Smart Contract Manager enforces certain contractual parameters of the data sharing agreement between a data provider and a consumer using predefined smart contracts, which are based on the legal agreements. The Smart Contract Manager component incorporates the conflict resolution, the explicit user-consent, and the auditable accounting component.

Smart wallet:
i3-MARKET wallets are key components that enable interaction between the different stakeholders and services in the i3-MARKET ecosystem. A wallet just stores and uses cryptographic material that, in i3-MARKET, are used to achieve the following features: authentication and authorization (by proving ownership of DIDs and Verifiable Credentials), and non-repudiation of data exchange (by digitally signing different operations).

The smart wallet building block is designed to be secure and user friendly and in a way that existing technologies can be easily added as i3-MARKET-enabled wallets. In this project, we are going to integrate three types of wallets:

- HW wallet: Hardware wallets are in charge of storing the user's private keys using a physical device as storage. It performs cryptographic operations to reduce the key exposition. This specific wallet satisfies the highest security policies since it is based on 'something you have' and you are the owner of the data.
- SW wallet: Software wallets store the user's private keys on the storage of a general-purpose device (e.g., computer or smartphone). It combines the security polices of 'something you have' and 'something you know'. Despite no specific hardware is needed, the loss of the device might imply losing the keys if no backups are made.
- Cloud wallet: Cloud wallets store the user private keys on secure cloud databases. Even though it has less strict security policies, it can offer much more functionalities than the other wallet subsystems, such as easier access and simpler key recovery.

Data monetization:
The data monetization building block provides a crypto currency solution that allows instant currency exchange among the participating data spaces/marketplaces, and also supports full audibility of all transactions. This

is vital for a fully decentralized solution, as it provides the basis for building trust in the federation backplane. The payment mechanism shall support micro-payments and requires minimal cost.

The standard payment component permits in advance an *a posteriori* payment for a specific dataset or piece of data with traditional payment systems, ensuring trust, security, and full auditability of data transfers through an *ad-hoc* non-repudiable protocol.

The tokenization component provides the creation of a crypto token for instant currency exchange among the participating data spaces/marketplaces ensuring the full auditability of payment transactions provided by the blockchain technology.

The micro payment component provides a mechanism for reducing the cost of crypto payment transactions especially for small amounts of tokens.

Semantic engine system – general description:
One of the core pillars in i3-MARKET is the semantic engine, which plays an important role in terms of registration and querying the offerings in a distributed and interoperable way. Semantic engine exposes different API endpoints for various tasks, for example, registration and querying as shown in Figure 1.5.

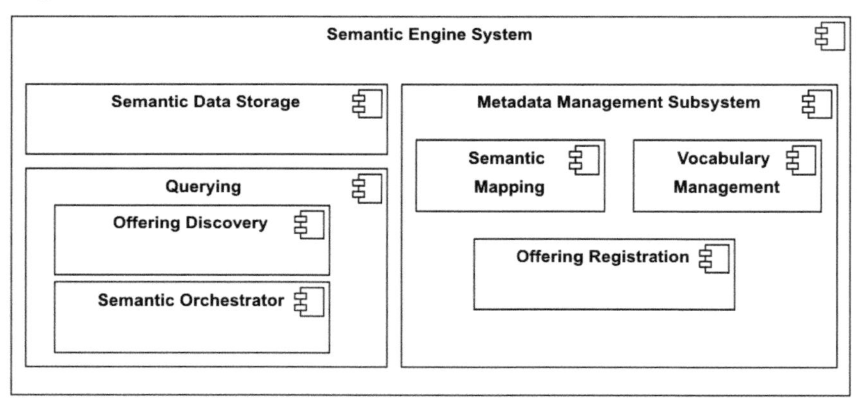

Figure 1.5 Semantic engine component diagram.

Inner building blocks:

Metadata management:

The metadata management subsystem encompasses three components:

- Offering registration: This component allows users to register offerings. Semantic engine exposes different endpoints for offering registration.

Examples are: (i) register data provider, (ii) register offering of a data provider, and (iii) update offerings.

- Semantic mapping: This component does semantic mappings and transforms (JSON to RDF) data received from API endpoints.
- Vocabulary management: This component keeps and manages all the vocabularies, defined as i3-semantic model, used in different components of the semantic engine.

Semantic storage:

This component communicates with the RDF triple store and pushes and retrieves data from that store.

In i3-MARKET, open-source virtuoso was tested as a triple store, which allows us to store RDF data and query using SPARQL query language. In general, triple stores are used to management tool for metadata, using semantic web query language (SPARQL) for accessing the information, which allows us to store RDF data and query using. At the same time, MongoDB is used to store metadata together with data descriptions and uses MongoDB query language. Figure 1.6 shows the data storage system.

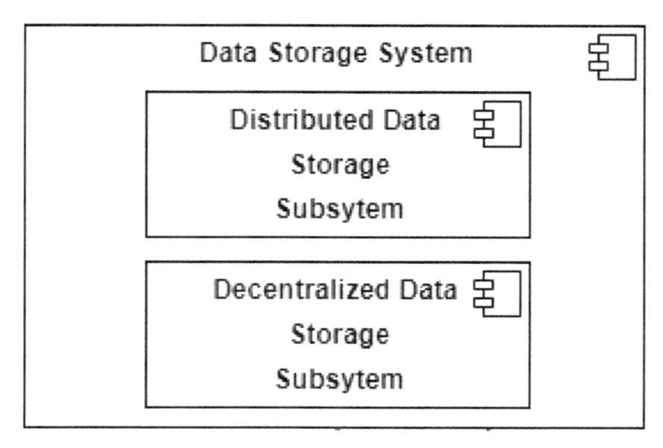

Figure 1.6 Data storage system component diagram.

Querying:

Two main components will be provided:

- Offering discovery: This component allows users, either data provider or consumers, to query already registered offerings.
- Semantic orchestrator: The role of this component is to manage synchronization with the distributed data storage component and the query processing, for instance, local or distributed query.

Data storage system – general description:

Inner building blocks:
The storage system consists of two main subsystems for implementing, respectively, the decentralized storage and distributed storage features, as shown in Figure 1.7. The subsystems are, at least in the initial architecture, relatively independent of other systems and, also, independent of each other.

Decentralized storage:
The diagram of decentralized storage subsystem is shown in Figure 1.7. The decentralized storage subsystem is implemented as a blockchain-based distributed ledger network. The software implementation is Hyperledger BESU in a permissioned setup using IBFT 2.0 consensus [10]. Hyperledger BESU uses internally an embedded RocksDB instance for storing linked blocks (the journal of transaction) and world state (the ledger). Hyperledger BESU can instantiate and execute smart contracts for supporting the use cases of the i3-MARKET framework.

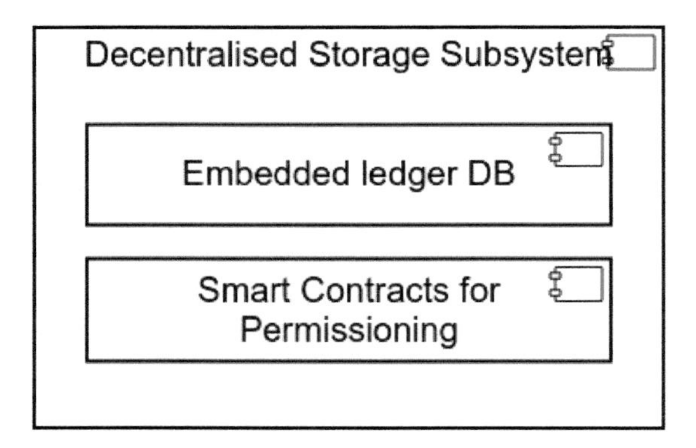

Figure 1.7 Decentralized storage component diagram.

Distributed storage:
The diagram of the distributed storage subsystem is shown in Figure 1.8. The subsystem consists of a distributed cluster of database nodes and an optional interface layer (will not be implemented for V1). The database provides an SQL interface to other i3-MARKET framework components. The software implementation of the internal database is CockroachDB that can be accessed via PostgreSQL-compatible wire protocol for which a large number of client libraries exist for different languages and platforms. CockroachDB

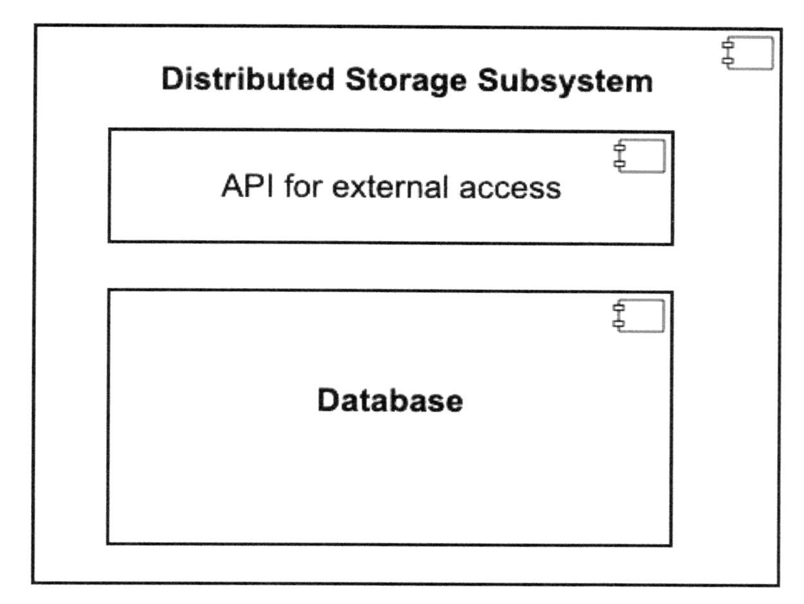

Figure 1.8 Distributed storage component diagram.

is a highly scalable and resilient distributed database. Only secure access to the database will be enabled; hence, all clients need to use private keys and valid certificates to access the database.

Deployment view:

The deployment or physical view "describes the mapping(s) of the software onto the hardware and reflects its distributed aspect"; the i3-MARKET deployment view is depicted in Figure 1.9. Four nodes constituted the i3-MARKET R1 cluster. On each node, it will be deployed a Backplane Gateway System and an instance of all the rest i3-MARKET main building blocks (trust, security, and privacy system, storage system, and data access system) giving backend support to the Backplane Gateway System. In addition to that, node 4 will host all the components related to the Semantic Engine Building Block in the form of free Open Source Software Tools for SMEs, developers, and large industries building/enhancing their data marketplaces. Figure 1.9 shows the proposed deployment as explained.

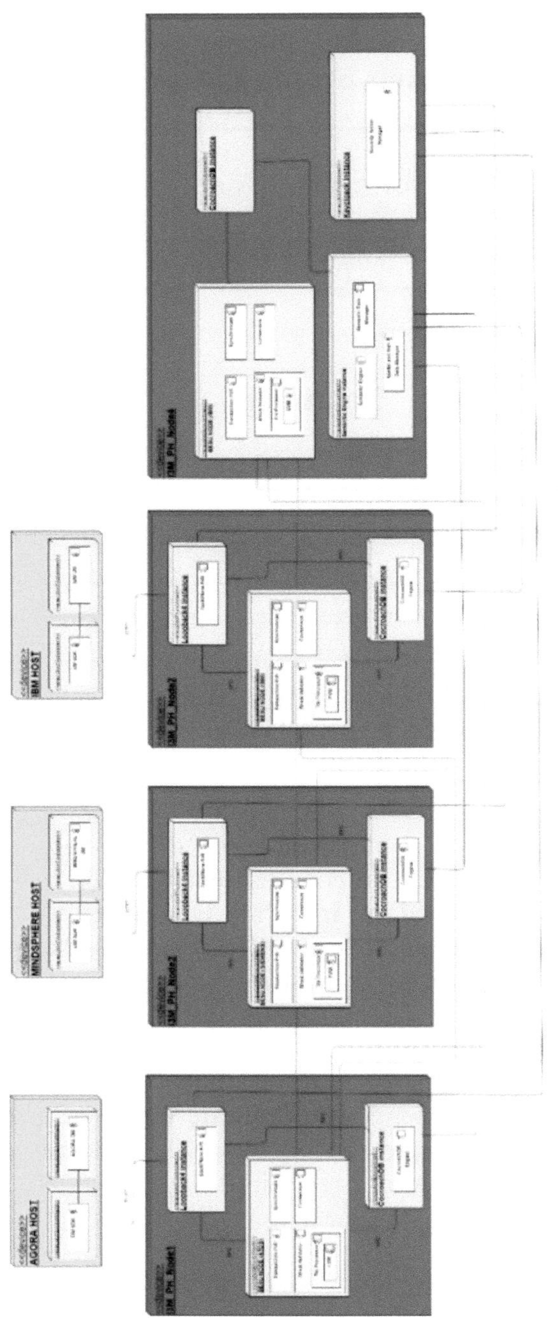

Figure 1.9 i3-MARKET deployment view.

2

Wallets and Smart Contracts

This chapter is focused on the specification and development of the i3M-Wallet, Auditable Accounting, Conflict Resolution, Explicit Consent, and Smart Contract Manager subsystems.

All the subsystem development is already public in the i3-MARKET GitHub and Gitlab repositories. For detailed information on every subsystem, one can jump to their specific sections in this book or check the documentation in the public repositories, which is constantly updated.

Table 2.1 summarizes the main technical contributions of the different building blocks addressed in this book: the i3M-Wallet, the Auditable Accounting, and the Smart Contract Manager.

2.1 i3-MARKET Wallet

There is a considerable amount of wallet applications. Some popular examples are MetaMask [6], TrustWallet [7], Exodus [8], or Electrum [9]. These applications use a dedicated app for iOS and Android, and browser extensions for desktop computers. Most of them are cryptocurrency wallets and are therefore targeted to operate with crypto tokens/currencies, showing balances, and allowing token transactions and swapping. However, there are not so many solutions facing the secure storage of W3C Verifiable Credentials and the selective disclosure of claims. Among them, the most common is to build the Wallet upon an SSI solution based on Sovrin [10] [11] or directly Hyperledger Aries [12] [13]. Fewer options have been found that use the Ethereum DLT, namely uPort [14], which has been discontinued, and Twala [15].

In i3-MARKET, the Wallet App has inherited some strong requirements:

- The technology must be open-source.
- It must work from the very beginning with Ethereum-like DLTs, as it is the case of Hyperledger BESU, the chosen DLT for i3-MARKET.

Table 2.1 Main technical contributions.

Building block	Main technical contributions
i3M-Wallet	High-level functionalities supporting the main i3-MARKET flows: • Authentication/authorization • Non-repudiation Protocol • Explicit Data-Owner Consent • Smart Contract Manager Complete SSI (DIDs, VCs, and selective disclosure) Wallet running on Ethereum-like DLTs with a complete open-source codebase and not locked in by any vendor infrastructure. Designed to also support other DLTs. Innovative, more secure, and privacy-preserving interface with the Wallet application, including a secure pairing protocol that does not require any external infrastructure. Designed to be able to integrate any crypto key wallet, including hardware wallets.[1] Nowadays, IDEMIA's hardware wallet is already integrated. Secure cloud vault allowing to completely operate (restore) the Wallet from any device without loss of context data.1
Auditable Accounting	Data is registered in a distributed high-availability database distributed storage. Use of a reliable, fast, scalable solution based on the use of DLTs and Merkle trees to reliably notarize the registered data.
Conflict Resolution/Non-repudiation Protocol	i3-MARKET is enforcing a fair cryptographically verifiable billing system with any kind of money, including fiat money. Reliable non-repudiable and cryptographically verifiable log of every data exchange. The logs are designed to not leak any sensitive data but to provide non-repudiable proof of a digital data exchange under the specification of a given data sharing agreement. These proofs can be used to support fair unfakeable billing with fiat or crypto money and also to support claims for eventual conflicts in the data exchange. In many cases, i3-MARKET Backplane can automatically solve conflicts based on these proofs.
Explicit Data-Owner Consent	i3-MARKET is, to the best of our knowledge, the only technology that enforces the existence of Explicit Data-Owner Consents when a provider is selling data. Data owners can at any time revoke the consent and their data will not be distributed any longer.

Table 2.1 *Continued.*

Building block	Main technical contributions
Smart Contract Manager	Data sharing agreements are modelled with coloured Petri nets, allowing their formal verification before they are translated to smart contracts. Smart contracts are developed using DAML, which make the development DLT agnostic and allows for translating our smart contracts to multiple DLTs, including i3-MARKET's Hyperledger BESU.

Obviously, it should be designed to be extendable for other DLTs in the future, with special focus on Hyperledger Aries.

- It must be able to integrate existing key wallets, such as IDEMIA's hardware wallet, for signing.
- It must support SSI flows, but also crypto tokens, as i3-MARKET is creating a custom one. Therefore, i3M-Wallet is going to be a hybrid wallet, supporting SSI and cryptocurrencies.
- Adoption of SSI technologies should be easy for the end-users:
    - The Wallet should be easy to backup and restore with no loss of information.
    - The end-user wallet should be universally accessible from any device (desktop computer, mobile phone, etc.).

In the beginning, as a short-term solution for implementing SSI, we adopted the uPort Wallet [14] since it was already working with Ethereum, supports the disclosure of verifiable claims, its codebase is open-source, and was mature enough and well-tested. Twala [15] and other Ethereum-based SSI solutions were in a very early stage in the beginning of i3-MARKET, with little to no support for implementing the issuance of Verifiable Credentials and the server part of a selective disclosure.

The uPort solution has now been split into two different projects, Serto [16] and Veramo [17], being just the second in the libraries for creating and managing DIDs and Verifiable Credentials without worrying about interop and vendor lock-in. Veramo is at the very core of i3M-Wallet and the identity solution of i3-MARKET.

In any case, it was clear that i3-MARKET success would need a custom Wallet App supporting at the same time:

- Ethereum-based DLTs such as Hyperledger BESU, and potentially others (currently analysing Hyperledger Aries).

- Management of digital identities and Verifiable Credentials, including selective disclosure of claims, as well as cryptocurrencies.
- Complete open-source codebase with a technological solution not locked in by any vendor infrastructure.
- Enhanced usability in the sense of being universally accessible/recoverable from any device.
- The integration of any key wallet into the App, including hardware wallets.

Besides all the above features, the current version of i3M-Walllet is innovative in terms of the following:

- The desktop application has been built with privacy and security as a main design goal. For that reason, i3M-Wallet is not a browser extension and runs as a multi-platform application that is securely paired to local applications (such as JavaScript running in a browser). Not sharing a process with the browser, as extensions do, prevents a bunch of potential "speculative execution" attacks and minimizes the exposure of the app to attacks performed by malicious websites. Moreover, running the wallet as an external application is more privacy-respecting for the end-user since the Wallet will not have any access to the data exchanged with a visited page, as it is partially the case with extensions, especially when using Firefox, which has got a more limited sandbox for extensions than Chrome.
- Even though there are complete and mature implementations of the selective disclosure of verifiable claims running on Hyperledger Aries and Sovrin, the solutions using Ethereum-like DLTs do not currently implement a complete selective disclosure flow. The closest solution was the popular but now abandoned uPort [14], which implemented a selective disclosure where users could agree or not to disclose a set of claims, but not to deal individually with each of them. The Serto [16] solution (derived from uPort) is aimed at providing that, but it is still not available for public testing. Twala [15] is more dedicated to digital signatures, with a somehow limited selective disclosure of identities' claims that, in any case, relies on the Twala ecosystem and closed infrastructure. The selective disclosure flows of i3-MARKET identity system, including the i3M-Wallet, is to the best of our knowledge the first complete selective disclosure flow on Ethereum-like DLTs that is completely open-source and not locked in by any vendor.

- The Wallet integrates a secure backup system (currently in testing phase) designed to not be tied to any vendor infrastructure. The backup is complete not only in terms of restoring cryptographic material but also to restore high-level i3-MARKET data, including identities, Verifiable Credentials, and non-repudiation proofs of data exchanges.

2.2 Auditable Accounting

Marketplaces need to record, audit, and provide availability and non-repudiation for data involved in exchanges. The auditing tasks in these systems are typically performed by a trusted third-party auditor (TTPA) who is responsible for checking the integrity of the content and thus for increasing stakeholders' trust in data exchanges. It is a centralized model where all the power and responsibility fall on the TTPA, which is a single point of failure and cannot be disputed by users. Decentralized architectures and protocols appear as an alternative to avoid those risks while providing the same quality of service (QoS).

The challenges of designing a feasible storage auditing framework emanate from the security challenges of decentralized solutions and the performance overhead due to on-chain operations. In this context, the work in [18] presents a solution based on a Merkle hash tree for Auditable Accounting. Their approach is similar to ours and the project is also open-source. However, the implementation is made for the Bitcoin network while our project uses an Ethereum-based network built with the Hyperledger BESU client. In other works, polynomial commitment schemes are used to create succinct proofs of data possession and guarantee data availability [19] [20] [21] [22]. However, i3-MARKET's Auditable Accounting system is faster and more scalable but with less complexity than the mentioned alternatives. This is made possible thanks to the following:

- the usage of Merkle trees for aggregating notarization proofs of the data to register;
- the use of a smart contract for storing just the roots of those Merkle trees on the blockchain, which allows for reliable verification while heavily reducing the needs of storage in the blockchain.

The storage of the registered data with their corresponding Merkle proofs (needed for verifying against the Merkle roots) is available and ready to use as a fully distributed database provided in the i3-MARKET ecosystem.

2.3 Conflict Resolution/Non-Repudiation Protocol

One of the main issues with digital data trading is related to the legal support if either the consumer or the provider does not adhere to the signed agreement. If not under the umbrella of a big player that assumes the risks, this situation diminishes the confidence in the data exchange and prevents the ignition of an ecosystem of digital data trading.

i3-MARKET provides a technology that relies on the use of a blockchain/DLT to build confidence in digital data trading. Contrarily to other approaches that also use a DLT for that purpose [23] [24] [25], i3-MARKET does not want to force its stakeholders to use specific crypto currencies/tokens (although it provides one if desired), which can be used to automatize payments when certain conditions are met; *i3-MARKET wants to build confidence on data exchanges with any payment system*, including the most common one: fiat money. As a result, i3-MARKET is, to the best of our knowledge, the first technology that uses a DLT just as a reliable ledger of the data exchange with the goal of supporting a *fair billing system* (also with fiat money) and to be able to *solve eventual disputes*.

The i3-MARKET innovative approach generates proofs of every data exchange that can be later used to prove what was exchanged, when it was exchanged, and under which data sharing agreement. i3-MARKET does not define per-se a payment system (although it provides a crypto token if desired) but generates cryptographically verifiable and reliable data that can be used to properly invoice, and to support eventual future disputes, since both the consumer and the provider, and any third party they allowed as well, can verify them.

Besides those disputes based on subjective opinions (for instance, a consumer not liking the acquired dataset), i3-MARKET can automatically solve disputes and even enforce penalties if it were part of the agreement.

2.4 Explicit Consent

i3-MARKET's architecture has been designed to allow all the stakeholders – namely providers, consumers, data owners, and marketplace operators – to meet the strictest policies in terms of privacy and data protection, which in fact leads to meet the GDPR requirements with little effort.

Article 4 of the GDPR [26] defines consent as "any freely given, specific, informed and unambiguous indication of the data subject's wishes by which he or she, by a statement or by a clear affirmative action, signifies agreement to the processing of personal data relating to him or her". Data controllers shall be able to demonstrate that they hold the Explicit Consent of the data subjects to process (Article 7) and/or trade their data. To the best of our knowledge, no technology is enforcing user consent to the point of preventing trading without it.

It is a remarkably innovative feature of the i3-MARKET project that the Explicit Consent of the data subjects is absolutely required for trading users' data.

2.5 Smart Contract Manager

There is noticeable interest in the literature related to automating service-level agreements, specifically data sharing agreements, by leveraging a distributed ledger technology (DLT), such as the blockchain. DLTs, and in particular smart contracts, help provide potential decentralized markets that furnish a peer-to-peer interaction between the different parties without third-party interference. Hence, this contributes to empowering the shared economy applications.

A framework that enforces the parameters of the legal data-sharing agreements with the use of smart contracts is proposed in [23]. As they describe, these parameters are automatically enforced. Moreover, they have a voting-based system. This voting system is external and acts as Conflict Resolution in case of any breaches to the agreement terms. However, these smart contracts are written in Solidity, which need to be formally verified and prove that they are error prone.

The authors of [24] present an approach based on blockchain and smart contracts to enable dynamic payments during the entire SLA lifetime (compensation value). A smart contract is implemented to detect and record any violations on the blockchain. Once the violation is detected via the "monitoring"-smart contract, the compensation value will automatically be transferred to the customer. In their work, they also use Solidity contracts.

Ocean Protocol, presented in [25], has proposed a new approach called service execution agreement (SEA), which brings the idea of SLAs to the blockchain. An SEA represents the service-level specification of an SLA, which can be translated into a smart contract. SEAs are implemented as smart contracts running on the blockchain. They have a modular design consisting of three parts: service identifier, conditions and fulfilment, and reward logic. Nevertheless, Ocean Protocol uses Solidity smart contracts running on Ethereum. Moreover, Ocean Protocol has integrated a reward logic into SEA components to reward a network of verifiers for their work. According to Ocean Protocol, the role of verifiers is to maintain data integrity and availability. However, this would require the interference of a third party (representing the network of verifiers).

In this work, we have modelled all the possible execution paths of the data sharing agreement (service-level agreement for the data market domain) using coloured Petri nets [27], a modelling method that allows describing a variety of resource types and execution logic in a way that can be formally verified. As a result, we can formally verify the modelled agreement's behavioural properties and then, with a clear understanding of how these contractual agreements are executed, translate them to smart contracts retaining the correctness and completeness of the modelled agreement. On top of that, unlike many of the presented work that relies on Solidity and Ethereum blockchain, we use the digital asset modelling language (DAML) [28], an open-source smart contracts programming language inspired by Haskell, which helps make our approach more general and focus more on the business logic and the design of our approach. DAML allows platform-independence and can be later integrated into several DLTs, including i3-MARKET's Hyperledger BESU.

3

i3-MARKET Wallets

3.1 Objectives

i3M-Wallets is a set of technologies that facilitate the interaction of the different i3-MARKET stakeholders with the Backplane API. It manages i3-MARKET identities, data agreements (with signature verification/generation), non-repudiation proofs (generation/verification), and secrets for data encryption/decryption.

All the code has been made publicly available at the i3M-Wallet monorepo [2]. Several packages are provided in this repo, but a standard i3-MARKET user/developer is likely only needing:

- **i3M Server Wallet:** It is an interactionless wallet implementation not requiring any user interaction. It has been designed to be operated by a "machine" or service. Current implementation is in TypeScript/JavaScript and can be easily imported to any JS project with NPM/Yarn.
- **i3M-Wallet Desktop App:** It is a desktop application (Windows, MacOS, and Linux) thought to be operated by end-users. The app can be securely paired to any application, allowing the application to interact with the wallet through an HTTP API. Wallet actions requested by any application will require explicit confirmation of the end-user through the app interface (window).
- **i3M-Wallet Protocol API:** A TypeScript/JavaScript library that can be used to easily connect to an i3M-Wallet Desktop App. It wraps all the functionalities provided by the wallet's HTTP API into convenient class methods. It works in Node.js (both ESM and CJS) and browsers. Follow the pairing example to properly pair your JS application to the wallet and start using the Wallet API.
- **i3M-Wallet OpenAPI Specification:** In order to get a better understanding of what functionalities of the wallet are provided to paired

applications, a developer should analyse the i3M-Wallet OpenAPI Specification [39] or just visualize it online at editor.swagger.io[1].

The complete documentation of every package is provided in every package's README of the open-source public repositories.

3.2 Technical Requirements

i3M-Wallets implement the following requirements:

Identity management:
The wallet implements key functionalities for enabling the self-sovereign identity (SSI) solution of i3-MARKET. These functionalities are described below.

DID management:
The i3-MARKET identity subsystem heavily relies on the use of distributed identifiers (DID) [40]. The wallet should then be able to manage DIDs, specifically:

- **Create DID:** The wallet should be able to create a DID and the complementary cryptographic key for managing it. The keys must be securely stored/managed.
- **Present DID:** The wallet should be able to present a DID upon request and prove ownership of it.
- **Resolve DID:** The wallet should be able to retrieve the public data associated with a DID, including (but not limited to) the public keys associated with the DID.
- **Verify asset signature:** The wallet should be able to verify signatures using the DID of the signer.
- **Sign assets:** The wallet should be able to sign assets using the private key associated with a DID.
- **Deactivate DID:** The wallet should be able to deactivate a DID.

Verifiable Credentials management:
i3M-Wallets handle i3-MARKET identities, which, in the end, are DIDs and a set of Verifiable Credentials issued for those DIDs. A Verifiable Credential [41] is a tamper-evident credential that can be cryptographically verified and

[1]https://editor.swagger.io/?url=https://raw.githubusercontent.com/i3-MARKET-V3-Public-Repository/SP3-SCGBSSW-I3mWalletMonorepo/public/packages/wallet-desktop-openapi/openapi.yaml

stores claims about an identity issued by different entities. An example of a Verifiable Credential could be a university diploma issued to a student. The specific requirements implemented by the i3M-Wallets with regard to the management of Verifiable Credentials are:

- **Verify Verifiable Credential:** Verify the validity of a Verifiable Credential by verifying the signatures of the issuers.
- **Share Verifiable Credential:** Share an owned Verifiable Credential upon request and prove ownership.
- **Store Verifiable Credential:** Store Verifiable Credentials associated with owned identities.

Secure data exchanges:
For a secure data exchange to happen, the i3M-Wallet should support the management of the cryptographic material needed for the secure data exchange, including the storage and verification of the data sharing agreements, the verifiable proofs for the non-repudiation protocol, and the cryptographic material associated with every data exchange. Please refer to Chapter 14 for better understanding the flow of a secure data exchange.

Specifically, an i3M-Wallet implements the following requirements regarding secure data exchanges.

- **Store data sharing agreements:** Store data sharing agreements associated with one of the identities managed by the wallet. The data sharing agreements are verified (both schema and signatures) before they are stored. The cryptographic material associated with the agreement, including the freshly created agreement-specific keys are also verified and stored.
- **Sign data sharing agreements:** Sign a data sharing agreement using one of the owned identities.
- **Store non-repudiation proofs:** For the conflict-resolution system to work, the wallet should store non-repudiation proofs for every data exchange, which can later be used to unequivocally prove that the data exchange happened and what was exchanged. Non-repudiation proofs are verified before being stored.

3.3 Solution Design/Blocks

The development of i3M-Wallets is organized in different packages/modules providing different functionalities. All the packages have been made publicly available at the i3M-Wallet monorepo, namely:

- Base Wallet: base-wallet [42]
- SW Wallet: sw-wallet [43]
- BOK Wallet: bok-wallet [44]
- Wallet Desktop: wallet-desktop [45]
- Server Wallet: server-wallet [46]
- Wallet OpenAPI: wallet-desktop-openapi [47]
- Wallet Protocol: wallet-protocol [48]
- Wallet Protocol API: wallet-protocol-api [49]
- Wallet Protocol Utils: wallet-protocol-utils [50]

The *Base Wallet* [42] package is a high-level implementation of the i3M-Wallet functionalities. It internally uses a crypto wallet. For such a purpose, it uses the so-called KeyWallet interface, which currently has several implementations:

- **SW Wallet** [43]: A software implementation of a hierarchically deterministic wallet, which can be recomputed with a seed.
- **BOK Wallet** [44]: A software implementation of a wallet implemented as a bag of (independent) keys.
- **HW Wallet**: A package allowing the use of IDEMIA's i3-MARKET HW Wallet as the internal KeyWallet. The implementation of this package involves IDEMIA's proprietary code implementing the open-source KeyWallet interface. A video demonstrating how it is used with the i3M-Wallet Desktop application can be watched at [51].
- **Wallet Desktop** [45] is the i3M-Wallet Desktop application providing a secure and convenient user interface to the i3-MARKET Base Wallet. It can be defined as a cross-platform facility tool that eases the communication between a wallet (software or hardware) and the i3-MARKET SDK via an HTTP API. Furthermore, it provides some features like wallet synchronization using a secure cloud vault. It also has a user interface (UI) to display the information of the selected wallet and ask for user consent if any wallet operation needs it. The Figure 3.1 shows an example of the Wallet UI.

If the wallet is to be run by a service instead of an actual person, the *Server Wallet* [46] should be used instead, which is distributed as a JavaScript/TypeScript library that can be easily instantiated from server code.

Both the Server Wallet and the Wallet Desktop expose some functionalities that can be used programmatically from paired applications. The *Wallet OpenAPI* [47] defines the internal HTTP API that is exposed (more details in section of Interface Description). However, for security reasons, the API is

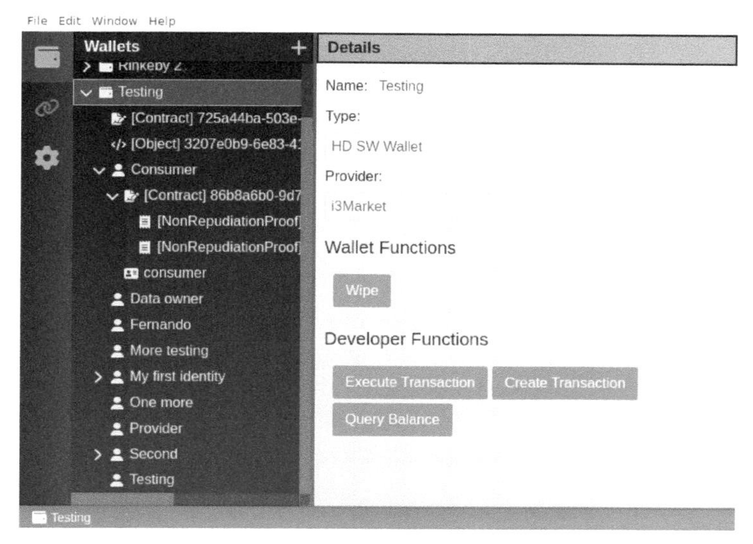

Figure 3.1 Wallet Desktop UI.

not directly available via HTTP. Indeed, it is encapsulated inside a secure session that is established after successful pairing of the wallet with an application.

Creating a secure session requires the execution of the *Wallet Protocol* [48], which implements the i3M-Wallet pairing protocol and the agreement of a secure (both encrypted and authenticated) session between the wallet and the paired application.

The Wallet Protocol enables any application to securely connect to the wallet. It solves two problems: the discovery of the wallet and the secure channel creation. The i3M-Wallet pairing protocol is designed to pair applications running in the same machine and to not require any external entity for the process, and it is heavily inspired in the Bluetooth Secure Simple Pairing protocol in use since Bluetooth 2.1 [52]. The protocol has been carefully designed and validated with a formal security analysis using Tamarin's prover [53], a tool that has also been used to validate TLS1.3, among other security protocols. The complete description and design are available in open access repositories.

In order to ease the use of the Wallet Protocol a set of libraries have been developed for both Node.js and browser JavaScript. *Wallet Protocol Utils* [50] defines a set of utilities for the pairing, including dialogs for setting the PIN in browser JS apps and Node.js, and session managers for properly managing wallet-protocol's sessions obtained after a successful pairing. Moreover, *Wallet Protocol API* [49] is a TypeScript/JavaScript library that wraps all the in-session encapsulated calls to the wallet's HTTP API into convenient class and methods.

3.4 Diagrams

Wallet Desktop:

Figure 3.3 shows the start-up flow of the Wallet Desktop. The first thing it does is loading the application configuration. It is a JSON file whose path depends on the operative system running the Wallet Desktop:
%APPDATA%\wallet-desktop\config.json on Windows
$XDG_CONFIG_HOME/wallet-desktop/config.json or ~/.config/wallet-desktop/config.json on Linux
~/Library/Application Support/wallet-dektop/config.json on macOS

Then the Wallet Desktop initializes the user interface and the so-called extra features, which nowadays is just an encrypted storage for supplementary data (cryptographic material is securely stored/managed by the KeyWallet). Figure 3.2 shows the UI asking for the encrypted storage password.

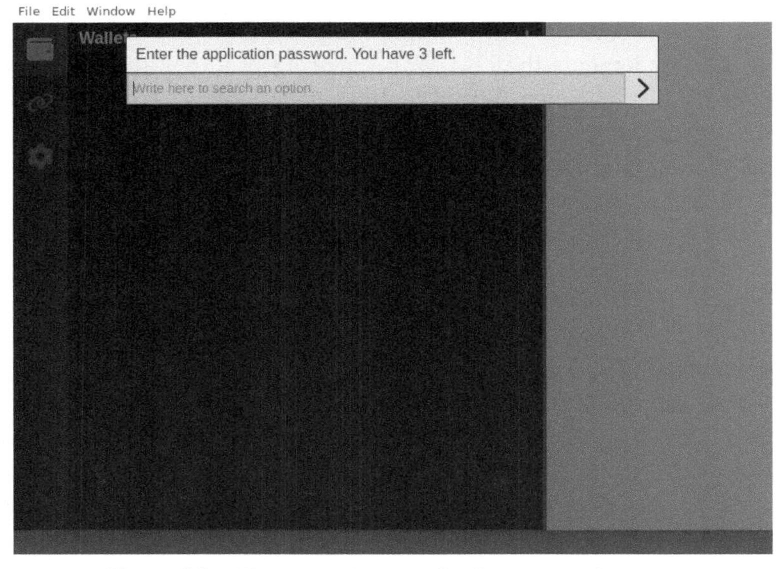

Figure 3.2 UI password request for the encrypted storage.

Figure 3.3 Wallet start-up flow.

The wallet factory oversees instantiating and the wallet modules. Wallet modules have an entry function that returns a wallet object. On the start-up, it will use the configuration file to select the current wallet.

Finally, it initializes the HTTP API. From now on, the application will listen to the user interface and the HTTP port in parallel; being securely encapsulated inside the HTTP connection, the wallet will receive calls to the internal HTTP API defined by the Wallet OpenAPI. The complete flow diagram is as defined in Figure 3.3.

As an example of invoking one of the wallet's API functionalities, let us follow the flow for generating a signature depicted in Figure 3.4.

As the Wallet Desktop receives an HTTP request in a Wallet Protocol's session, the request encapsulates (both encrypted and authenticated) an HTTP call to perform a signature; the OpenAPI Validation express module uses the OpenAPI definitions present in the package wallet-desktop-openapi to validate the request body. If the request is valid, the sign API handler gets the current wallet selected by the user and it calls its sign function. The current wallet can be any kind of hardware or software wallet supported by Wallet Desktop.

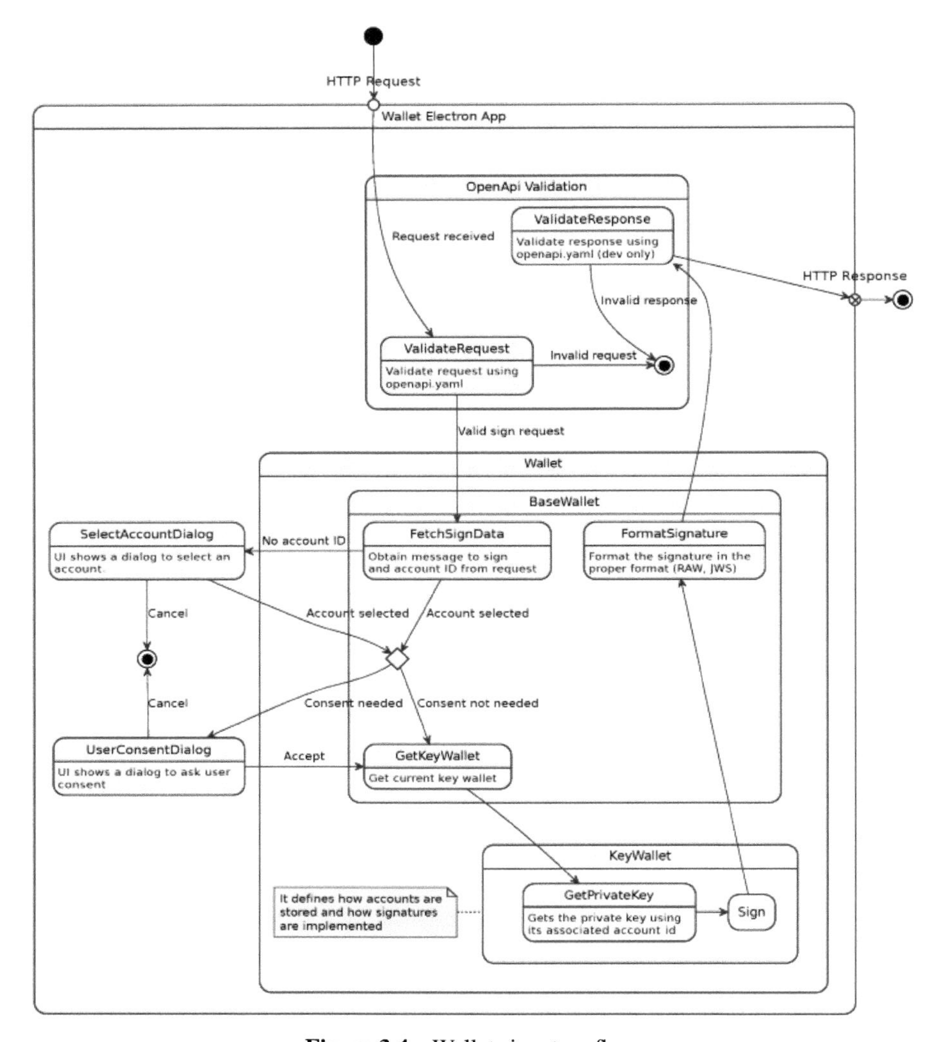

Figure 3.4 Wallet signature flow.

BaseWallet is a class present in the base-wallet package that offers a default implementation of a wallet. Nonetheless, it requires an object implementing the KeyWallet interface to work. KeyWallet defines the low-level implementation of the wallet: it can store keys and use them to sign. Note that by splitting the wallet in BaseWallet and KeyWallet, software wallet (SW) and hardware wallet (HW) can share high-level wallet functionalities, such as signature formatting or key recovering.

To perform a signature, an account must be selected. If the API request does not contain an account ID, the Wallet Desktop will show the list of accounts inside the wallet so that the user can select one. Then, the BaseWallet will check the application configuration to check if signatures need user consent. If true, the wallet application will show a dialog.

Once the account ID is retrieved and the user consents the signature, the KeyWallet is now able to perform a signature. First, it will get the private key associated with the current account ID, and then it will sign the requested message.

Since the signature format is a common functionality of any wallet, KeyWallets must return signatures in DER encoding so that the BaseWallet can format it in the requested format.

Finally, the Wallet Desktop application builds the response message. On development, the OpenApi Validation module uses the OpenApi definitions to verify the response format (Figure 3.4). If it is correct, the response is sent back to the application encapsulated inside the Wallet Protocol's secure session.

OIDC Authentication:

Wallet Desktop can be used in conjunction with the i3-MARKET OIDC provider to authenticate users. This flow is added here so that views of flows using the Wallet are in this book. However, more detailed information should be available at deliverables in "Trust, Security and Privacy Solutions for Securing Data Marketplaces" at https://www.i3-market.eu/research-and-technology-library/.

Figure 3.5 shows how to perform this authentication flow. As a summary, it specifies how the OIDC provider uses the i3-MARKET SDK to create a selective disclosure request asking a set of verifiable claims to the Wallet Desktop. If the user has them and accepts the disclosure, the Wallet Desktop will answer with a list of verifiable claims along with a proof of ownership. This proof consists of a signature of the disclosure response using the private key of the user DID.

The flow starts when an OpenID Connect relying party (OIDC RP) redirects a user to the OIDC provider using a scope. The scope is a string that specifies which verifiable claims are requested. The scope supported are:

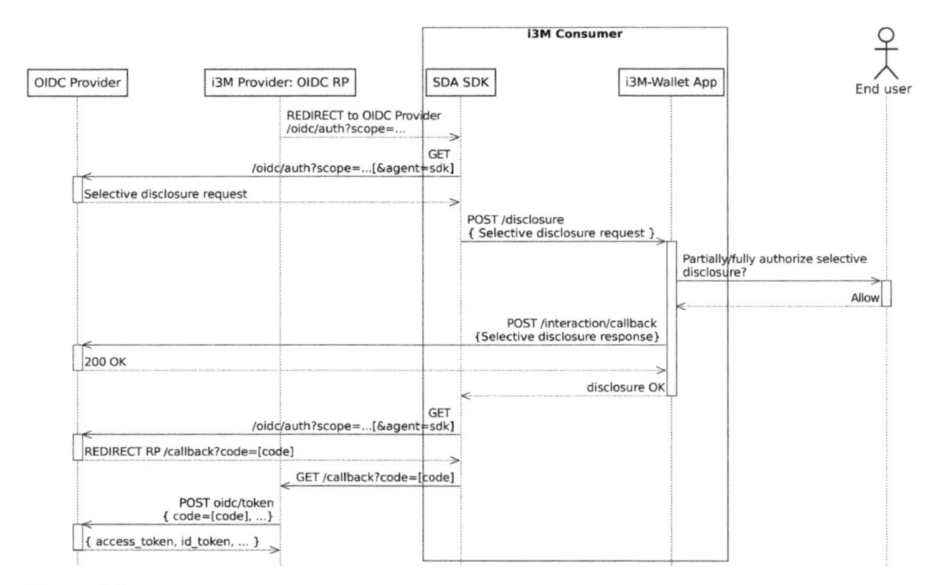

Figure 3.5 OIDC authentication using Wallet Desktop and the i3-MARKET SDK (sequence diagram).

- **openid:** The standard OpenID scope. It asks the OIDC provider to return an *idToken*, which is a jwt token containing the user information.
- **vc:** This scope is used to add the verifiable claims inside the *idToken*.
- **vc:*<claim>*:** It notifies to the OIDC provider that users may want to present an optional claim called *<claim>*. It is useful to ask optional claims like some extra user profile information.
- **vce:*<claim>*:** It notifies to the OIDC provider that users must present a valid claim called *<claim>* to proceed with the authentication. It can be used to ask users to present a claim that demonstrates their role (consumer, provider, data owner, etc.).

An example of scope is *"openid vc vce:consumer vc:profile"*. Using this scope, the OIDC provider builds and signs a selective disclosure request. It will be sent to the Wallet Desktop so as to obtain the required claims.

The Wallet Desktop HTTP API runs locally on the user computer. Cloud servers cannot access it directly so that the disclosure request must be sent by the user's computer. An approach is to create a simple frontend that only sends the selective disclosure request to the Wallet Desktop.

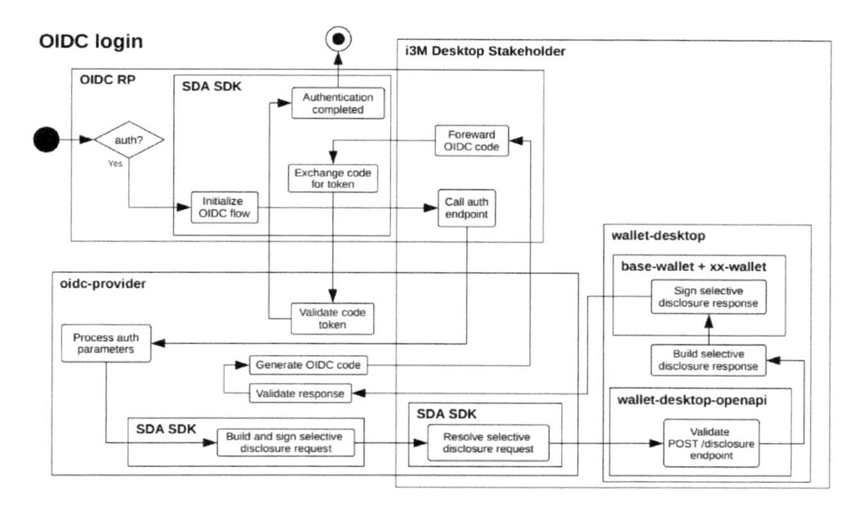

Figure 3.6 OIDC authentication using the wallet (activity diagram).

Then, the wallet displays a dialog to individually consent each verifiable claim. If the user allows the disclosure, the accepted verifiable claims will be sent back to the OIDC provider. After verifying all the claims, the OIDC provider will deliver an *idToken* and an *accessToken* to the i3M provider. Figure 3.6 is an equivalent diagram but putting more emphasis on where each block acts.

3.5 Interfaces

All the i3M-Wallet packages, but the i3M-Wallet Desktop App, are provided as TypeScript/JavaScript packages. The code is properly commented and TypeDoc has been used to convert comments in the TypeScript source code into rendered HTML documentation. As a result, the documentation is conveniently available when coding and also as HTML pages that can be accessed from the package's READMEs. As an example, Figure 3.7 shows a fragment of the API for the Server Wallet package.

The Wallet Desktop application follows the Wallet OpenAPI Specification. It is a REST-like API with four entities: identity, selective disclosure, resource, and transaction. It also has a set of helper functions.

- **Identity:** It can be used to create or list the DID of the user.

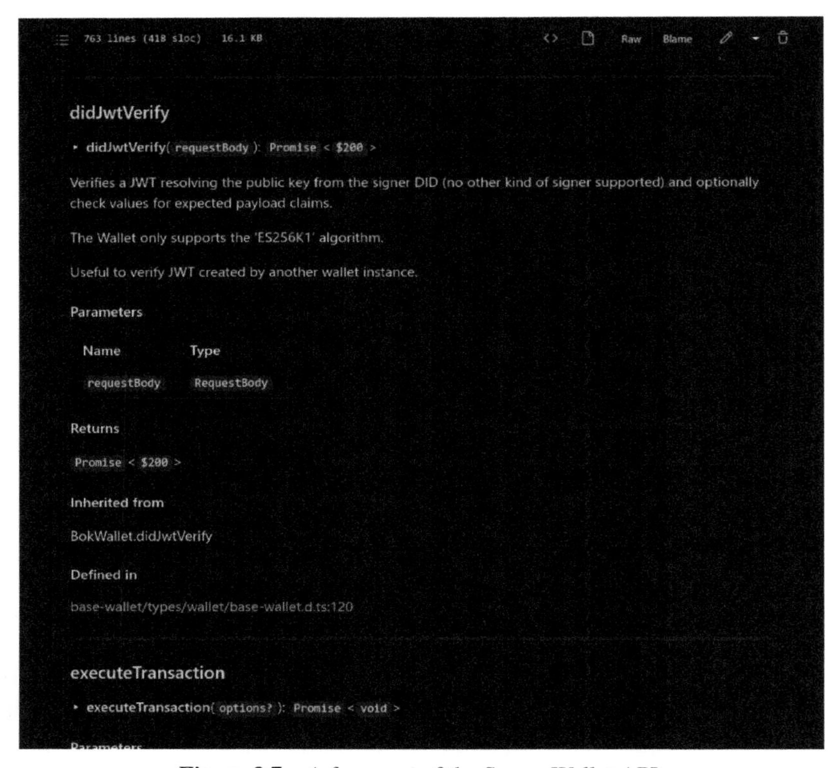

Figure 3.7 A fragment of the Server Wallet API.

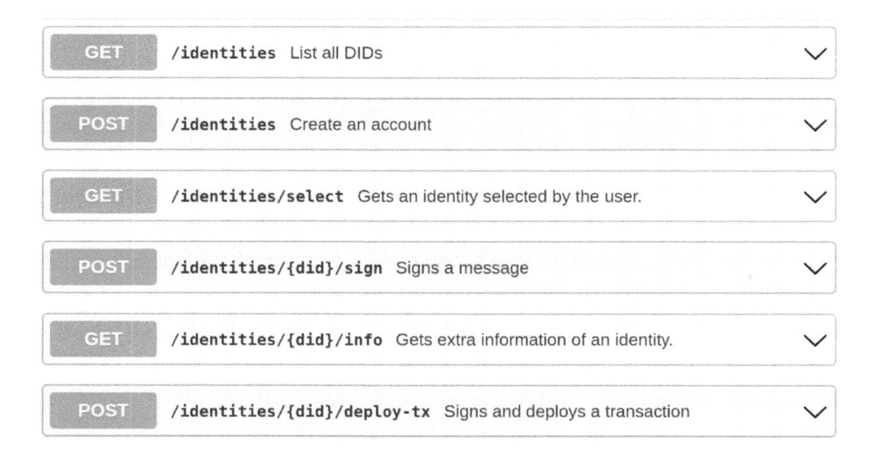

- **Selective disclosure:** Used to get a set of resources proving its ownership. They are signed with the requester private key.

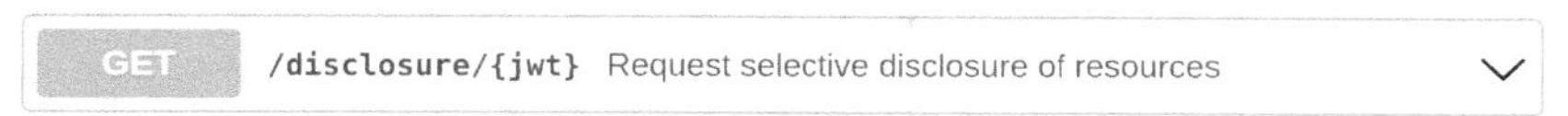

- **Resource:** Besides identities and secrets, the wallet *may* securely store arbitrary objects in a secure vault. The list of requests for resource is shown in the following diagram:

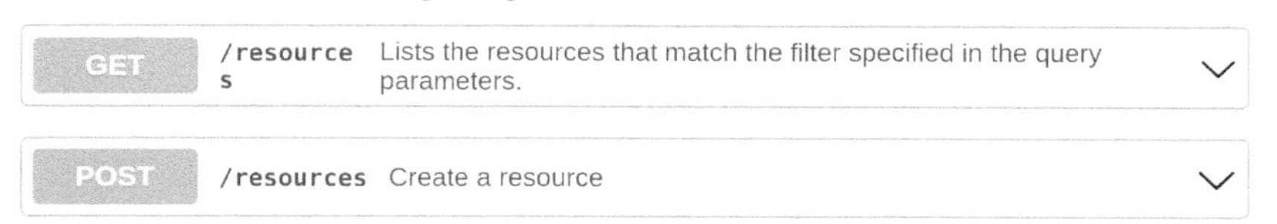

- **Transaction:** Endpoints for deploying signed transactions to the DLT the wallet is connected to.

- **Utils:** Additional helper functions.

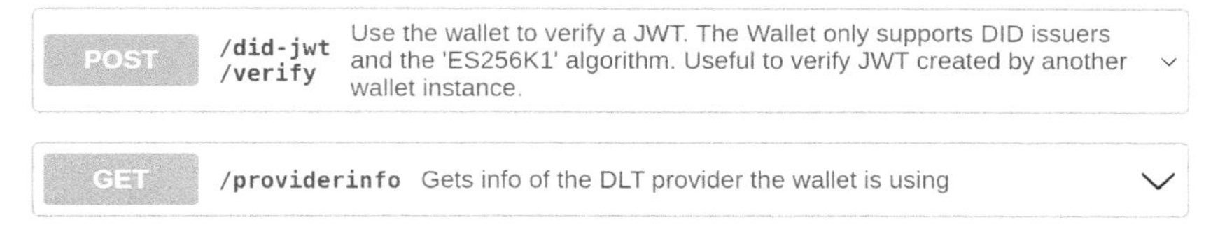

3.6 Background Technologies

The Wallet Desktop application uses Electron [54] to build a cross-platform application. Electron is a framework for creating native applications using web technologies using chromium. It also has implemented the Node.js core libraries so that the application can easily access to the operative system functionalities.

The libraries of the wallet monorepo use Ethers.js and Veramo to implement the integration of the Wallet with the i3-MARKET DLT:

– Ethers.js [55] is a complete and compact library for interacting with the Ethereum blockchain. It was originally designed for use with ethers.io and has since expanded into a more general-purpose library.

- Veramo [56] is a JavaScript Framework for Verifiable Data that was designed from the ground up to be flexible and modular, which makes it highly scalable. It can run on several environments: node, mobile, and browser. Its main utility is to make easy the use of DIDs, Verifiable Credentials, and data-centric protocols to bring next-generation features to users.

Wallet OpenAPI meets the OpenAPI specification [57]. Wallet Desktop validates all inputs against the OpenAPI schema using express-openapi-validator [58].

Documentation for the different packages has been made available, thanks to TypeDoc [59].

4

Auditable Accounting

4.1 Objectives

The auditable accounting component is responsible of registering auditable logs. As such, this component is one of the main tools to enhance the trust in the ecosystem of data marketplaces. Our solution must enforce the data sharing agreement terms, agreed upon all involved parties, by recording them in an auditable, transparent, and immutable way. Smart contracts are the key part of the proposed solution for auditable accounting. Figure 4.1 shows that the auditable accounting component is an abstraction layer to access the smart contracts and to allow the integration with the rest of the platform. The auditable accounting component is a service that includes an API to automate the process of logging and auditing interactions between components and record the registries in the blockchain.

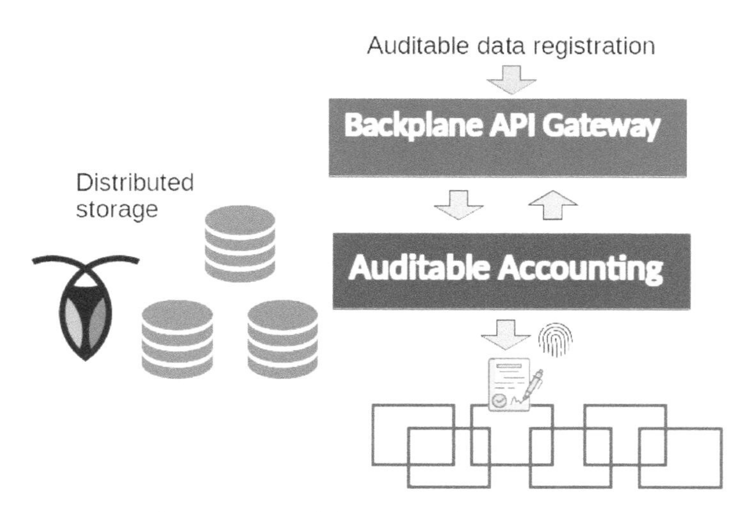

Figure 4.1 Auditable accounting architecture.

The auditable accounting development has been made publicly available in the i3-MARKET GitHub and Gitlab repositories (e.g., https://github.com/i3-Market-V3-Public-Repository/SP3-SCGBSSW-AA-AuditableAccounting). The Table 4.1 summarises the technical contributions used to design and implement the i3-MARKET Auditable accounting component.

4.2 Technical Requirements

Table 4.1 Main technical contributions.

Name	Description	Labels
Auditable Log	i3-MARKET needs to be able to log data and events in blockchain. It is a key component for accounting, billing, and conflict resolution. It is also important to control access to sensitive information and to detect potential data breaches. A public distributed ledger will be used to store non-repudiable and reliable proofs of the required actions. **Children:** 1. Auditable accounting: marketplace billing 2. Auditable accounting: conflict resolution 3. Auditable accounting: providing sensitive data 4. Providing sensitive data 5. Conflict resolution 6. Marketplace billing 7. Consumer billing 8. Provider billing **Parents:** 1. REQ-B-005 – i3-MARKET will ensure *trust* 2. REQ-B-008 – i3-MARKET will provide a payment solution 3. Semantic description of the SLS and the subscription	**V1** **Epic** **Data marketplace** **Data consumer** **Data provider** **Data owner**

4.3 Solution Design/Blocks

The solution must be scalable and cost-efficient. In this regard, transaction costs can be a considerable problem if, as it is expected, the number of auditable registries that need to be stored in the blockchain is high. To overcome this problem, it is a requirement to implement a transaction optimizer to efficiently register substantial amounts of data in the blockchain without incurring in excessive costs. To achieve this, we first store registries in an internal database of the component and then aggregate the registries with a Merkle tree to minimize the number of blockchain transactions and provide the appropriate data for proving each individual registry.

The smart contract managed by the auditable accounting component is used to store the necessary evidence of the aggregated registries from the DSA. Once the registration process is complete, the auditable accounting component will save a copy of all the information needed to verify that the registration was successful in the blockchain. This information can be consulted and obtained later by the marketplace users. The auditable accounting component provides the functionality to trace registries and obtain "certificates" of them that can be publicly or privately used to prove that a certain registry was performed. Users must be able to download these certificates and validate the registry without further interaction with the auditable accounting component having a proof that can be universally validated without the intervention of any other entity or software component. The certificate of a DSA will provide: the blockchain that has been used to create the auditable data registration, the address of the smart contract used, and the "proof of registry" of the associated data.

The auditable accounting component is a service that includes an API to automate the process of logging and auditing interactions between components and record the registries in the blockchain. As shown in Figure 4.1, in general, the API of the AA module is accessed through the Backplane API gateway. Additionally, the auditable accounting component can be accessed directly from any internal component of the platform.

On the other hand, to allow external parties to check that logs have been properly registered in the blockchain, interested parties need to obtain certain data from the distributed ledger as well as some off-chain data provided by the auditable accounting module via an API. This off-chain data are essentially Merkle proofs for each individual record. It is important that the off-chain data is provided with high availability. For this reason, the auditable accounting module uses the distributed storage component. In this way, high availability and data replication is provided to the relevant off-chain data required to store the registries and verify auditable logs.

Database model:

The database model proposed for this component is based on two SQL tables. The first one is the related one with the blockchain. It contains the transactions prepared or sent to the blockchain. Figure 4.2 shows the deployed columns as follows:

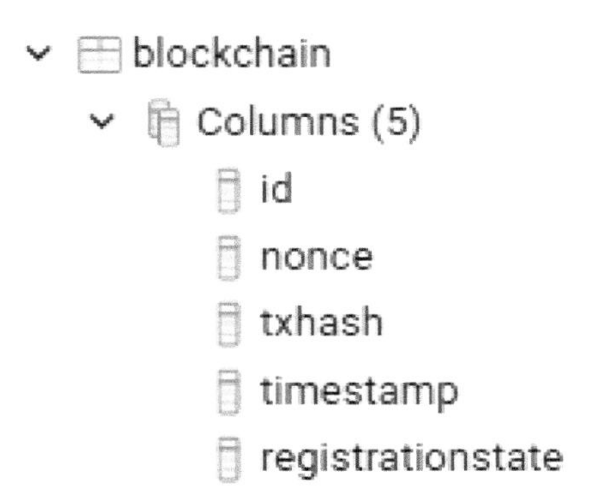

Figure 4.2 Auditable accounting library distribution.

- **Id:** Primary key to link with the other table.
- **Nonce:** Nonce from the account to build the transaction.
- **Txhash:** Hash of the transaction.
- **Timestamp:** Exact date of the creation of the transaction.
- **Registrationstate:** Status of the transaction.
    - **Unregistered:** Transaction not created.
    - **Pending:** Transaction created but not sent to the blockchain.
    - **Mined:** Transaction sent with less than 12 block confirmations.
    - **Confirmed:** Transaction sent with more that 12 block confirmations.

On the other hand, the registry table is responsible to store the proofs of the data hashes that want to be validated against the blockchain.

It contains the following columns:

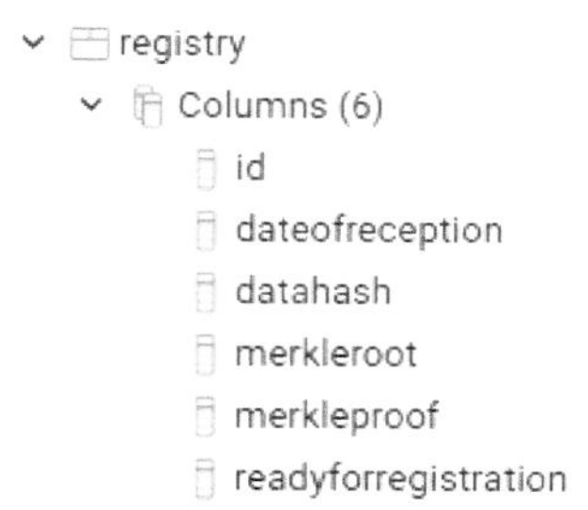

- **Id:** Primary key to link with the other table.
- **Dateofreception:** Date when the data is received.
- **Datahash:** Cryptographic hash function of the data. It is one of the leaves of the Merkle tree.
- **Merkleroot:** Root of the Merkle tree.
- **Merkleproof:** Concatenated hashes that allow to validate the datahash to the root of the tree.
- **Readyforregistration:** Boolean to indicate if the tree is built and ready to be deployed in the blockchain.

Smart contract:

The smart contract deployed for this component just stores the root of the Merkle tree that summarizes all the data hashes stored in the database. It only allows to modify that value by the owner of the smart contract, which shares the same account with the auditable accounting. Also, it includes the capability to subscribe to an event that notifies you about a new root released. The Solidity code is the following:

```
// SPDX-License-Identifier: MIT
pragma solidity >=0.4.22 <0.8.0;

contract AuditableAccounting {
  uint256 public currentRoot;
  address public owner;


  event newRegistry(uint256 prevRootHash, uint256 currentRootHash);

  modifier onlyOwner(){ require(msg.sender == owner, "sender must be the
contract owner"); _; }

  constructor() { owner = msg.sender; }

  function setNewRegistry(uint256 _newRoot) public onlyOwner {
    emit newRegistry(currentRoot, _newRoot);
    currentRoot = _newRoot;
  }
}
```

4.4 Diagrams

The workflow to register auditable data is shown in Figure 4.3.

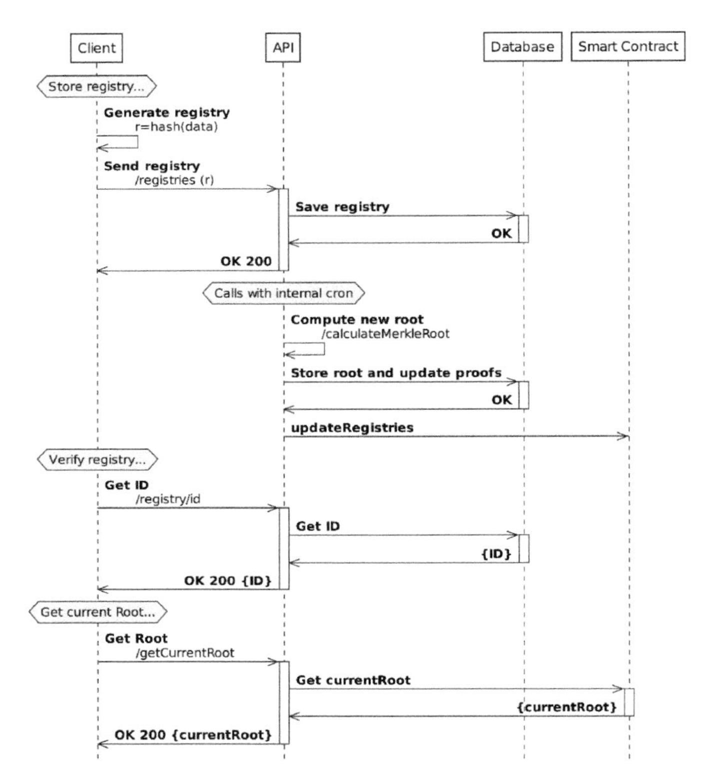

Figure 4.3 Auditable accounting flow.

The hash of the data to be registered is sent to the API using the endpoint/registries. Each hash to be registered is stored by the auditable accounting module in distributed storage. Then, the endpoint */calculateMerkleRoot* has to be called. When called, this endpoint creates the structure that is going to be registered in the blockchain. In more detail, this structure is a Merkle hash tree. The controller of the endpoint computes the Merkle root with all the pending registries, computes an individual proof for each registry, and stores these proofs in the distributed storage. Additionally, a transaction to be sent to the blockchain is created and stored in the blockchain SQL table in the distributed storage. Next, the endpoint */updateRegistries* can be called to store the Merkle root of the registries in the blockchain via the smart contract. We would like to stress that the endpoints */calculateMerkleRoot* and */updateRegistries* can be called with a "cron job" or similar to schedule registrations in the blockchain at the desired frequency. Finally, if a party wants to verify a certain registry, it can call the endpoint */registry/:id* to obtain the corresponding Merkle proof, compute the Merkle root from this proof, and compare it to the root registered in the smart contract. If both are the same, this means that the registry is valid.

4.5 Interfaces

The component is built from a Loopback 4 framework, which facilitates the management of the smart contract and the database generating an API that allows to integrate the procedures with the Backplane. But, as a high-level definition, the endpoints are divided into two controllers.
Firstly, the RegistryBlockchain controller manages the smart contract interactions and has the following endpoints:

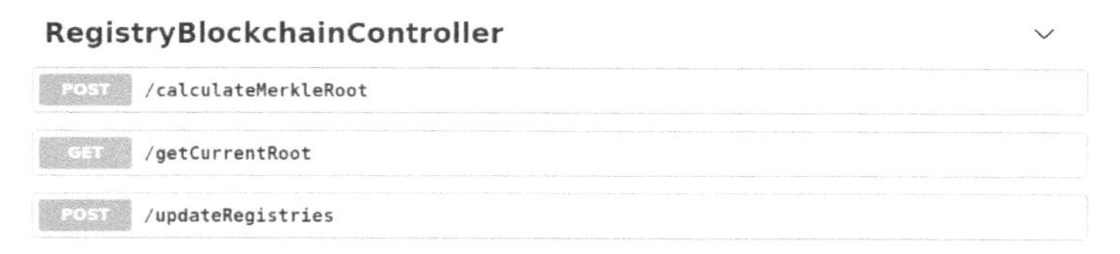

- **/calculateMerkleRoot:** Gets the pending registries from distributed storage that are not included in the current root and computes the new one.
- **/getCurrentRoot:** Gets the current root from the smart contract.
- **/updateRegistries:** Updates the status of the stored transactions and computes a new transaction.

On the other hand, there is the registry controller, which is responsible to manage the data hashes that are included in the auditable accounting system.

- GET **/registries/count:** Returns the number of stored registries.
- PUT **/registries/{id}:** Forces the creation of a specific registry.
- PATCH **/registries/{id}:** Updates a specific registry.
- GET **/registries/{id}:** Returns the value of a specific registry.
- DELETE **/registries/{id}:** Removes a specific registry.
- POST **/registries:** Generates a new registry.
- GET **/registries:** Returns the value of the registries.

4.6 Background Technologies

- **Solidity:**

Solidity is an object-oriented, high-level language for implementing smart contracts. Smart contracts are programs that govern the behaviour of accounts within the Ethereum state. It is a curly-bracket language. It is influenced by C++, Python, and JavaScript and is designed to target the Ethereum virtual machine (EVM).

Solidity is used to develop the smart contract deployed on the blockchain, which is responsible to store the root of the Merkle hash tree.

- **PostgreSQL:**

PostgreSQL is a powerful, open-source object-relational database system with over 30 years of active development that has earned it a strong reputation for reliability, feature robustness, and performance.

PostgreSQL is used to store the registries and the Merkle proofs of each registry.

- **Loopback** 4**:**

LoopBack 4 is an award-winning, highly extensible, open-source Node.js and TypeScript framework based on Express. It enables you to quickly create APIs and microservices composed from backend systems such as databases and SOAP or REST services. Also, it allows to manage custom data sources like a smart contract.

Loopback is used to generate the API that manages the registration of the data, the computation of the Merkle hash trees, and the smart contract executions.

5

Conflict Resolution/Non-repudiation Protocol

5.1 Objectives

The conflict resolution system's main goal is to prevent and/or solve conflicts when invoicing for a given exchange of data. It is therefore a core subsystem for the i3-MARKET secure data exchanges.

For the conflict resolution system to work, the i3-MARKET Non-repudiation Protocol (NRP) must be executed with every exchanged block of data. The Non-repudiation Protocol generates verifiable proofs of the data exchange that can be used to later prove that a given digital data exchange happened and that it met the agreed conditions (based on a data sharing agreement).

If the NRP is followed, the NRP proofs can be used to support fair unfakeable billing with fiat or crypto money and to prevent or solve eventual disputes with the data exchange alike.

A complementary conflict-resolver service (CRS) has been developed, which can be run by any trusted third party to issue verifiable signed resolutions regarding the execution of the NRP.

In short, as per the above explanation, the conflict resolution/Non-repudiation Protocol system relies on two subsystems, both already made publicly available in the i3-MARKET GitHub and Gitlab repo:

- the Non-repudiation Protocol library [60];
- the conflict-resolver service [61].

Updated detailed documentation can be found in, e.g., https://github.com/i3-Market-V3-Public-Repository/SP3-SCGBSSW-CR-Documentation#conflict-resolution--non-repudiation-protocol.

5.2 Technical Requirements

The conflict resolution/Non-repudiation Protocol must prevent the following situations between the two peers of a data exchange, namely provider and consumer:

- to deny that a given data-block exchange happened;
- or to assert that a data-block exchange that did not happen, happened.

As a result, providers will not be able to invoice a consumer for a data-block not exchanged; and consumers will not be able to deny or cancel a payment for a data-block that was successfully exchanged.

For it to happen, every block of data must be exchanged using the NRP. Accounted proofs give no room to alter the invoicing (fiat money) or the crypto payments (i3-MARKET tokens) if both entities reliably execute the protocol; otherwise, the conflict resolver service can be invoked to univocally solve which entity is intentionally or unintentionally malfunctioning.

5.3 Solution Design/Blocks

The Non-repudiation Protocol starts with a provider Alice, hereby A, sending a signed proof of origin (PoO) along with an encrypted block of data to a consumer Bob, hereby B.

An overview of the protocol is depicted in Figure 5.1, and more detailed sequence diagrams of every step are provided in the following sections.

After validating the PoO, B will demonstrate his will to get the data by sending a signed proof of reception (PoR). Just recall that B is at this point not yet able to decrypt the data since he does not know the secret to decrypt them.

The PoR is a proof that can be used by A to demonstrate that B is committed to get the secret to decrypt the block of data.

Now A can release the secret as part of a proof of publication (PoP). However, as B may state that he did not receive the PoP, A also publishes the secret to the ledger. It is now under B's responsibility to get the secret from the ledger since he implicitly agreed to it when sending the PoR.

For A to create a valid invoice for that block of data, she *must* present a valid PoR and demonstrate that the secret was published to the ledger within the agreed delay (part of the agreement). As a result, the lack of one or both proofs will result in an invalid invoice.

The conflict-resolver service (CRS) can be queried to provide a signed resolution about the Non-repudiation Protocol associated with an invoice

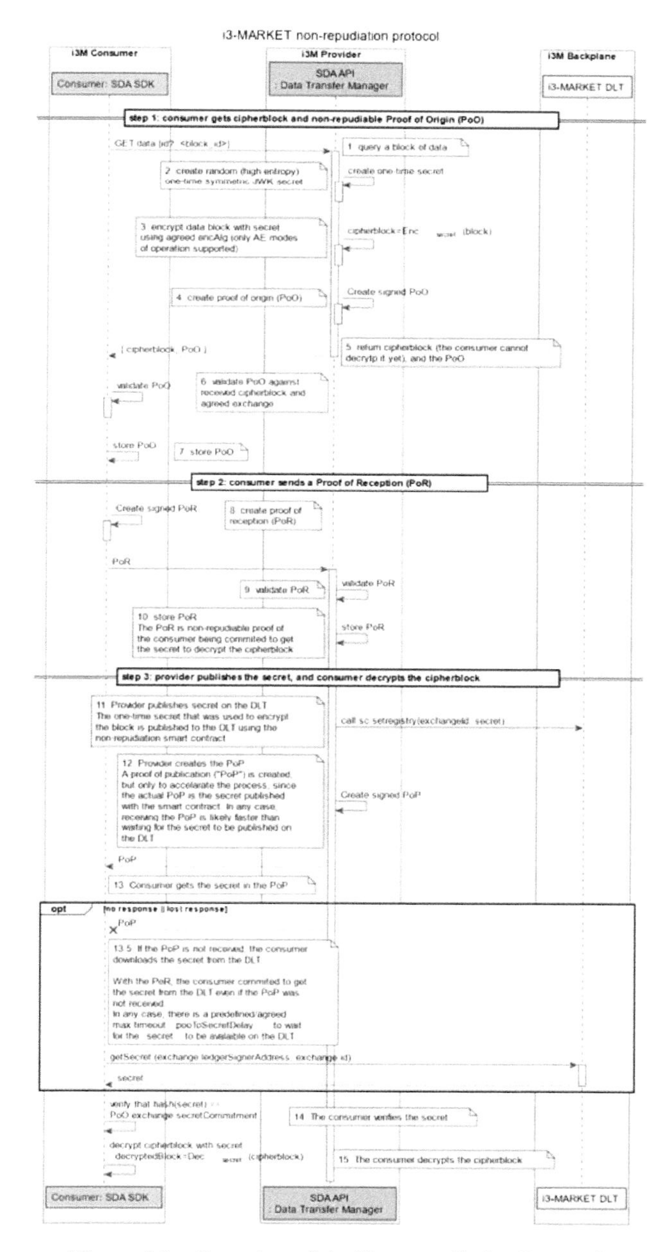

Figure 5.1 Overview of the Non-repudiation Protocol.

being valid or invalid. It could be invoked by either the consumer or the provider. The latter should be mandatory, being the resolution sent along with the invoice to the consumer.

However, this resolution does not ensure that the published secret could be used to decrypt the encrypted block of data. If the consumer B is not able to decrypt the cipherblock, he could initiate a dispute on the CRS. The CRS will also provide signed resolution of whether B is right or not.

5.4 Diagrams

This section presents detailed diagrams for the Non-repudiation Protocol and conflict resolution already depicted in the previous section. For a diagram with high-level overview of the NRP, please refer to Figures 5.2–5.6 for the different interactions and use cases.

- **NRP – step 1: consumer gets cipherblock and non-repudiable proof of origin (PoO).**
- **NRP – step 2: consumer sends a proof of reception (PoR).**
- **NRP – step 3: provider publishes the secret, and consumer decrypts the cipherblock.**
- **Conflict resolution: verification (NRP completeness).**
- **Conflict resolution: dispute.**

5.5 Interfaces

A standard i3-MARKET developer interacts with the conflict resolution/Non-repudiation Protocol system using the API of the non-repudiation library from the JavScript/TypeScript code or querying the conflict resolver service HTTP API.

API of the non-repudiation library:

The non-repudiation library API is a documented typescript library whose API can be properly documented "on the fly" while programming. Besides that, automated TypeDoc documentation is generated and available at https://github.com/i3-Market-V3-Public-Repository/SP3-SCGBSSW-CR-NonRepudiationLibrary/blob/public/docs/API.md.

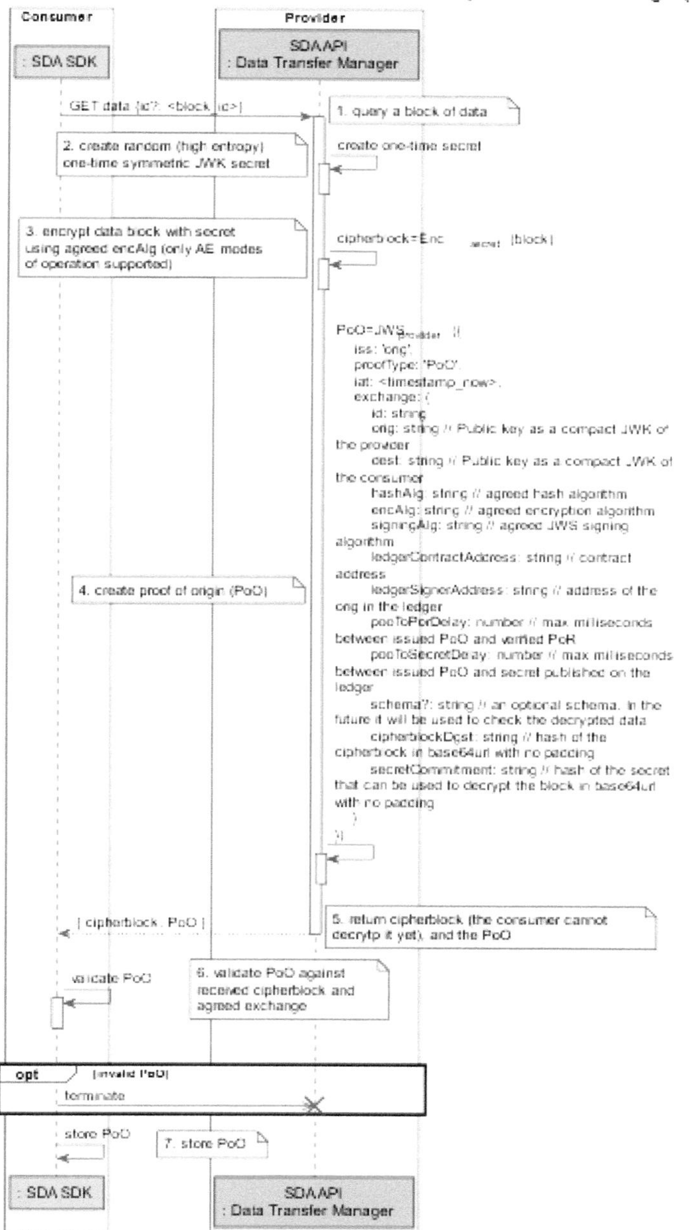

Figure 5.2 NRP – step 1: consumer gets cipherblock and non-repudiable proof of origin (PoO).

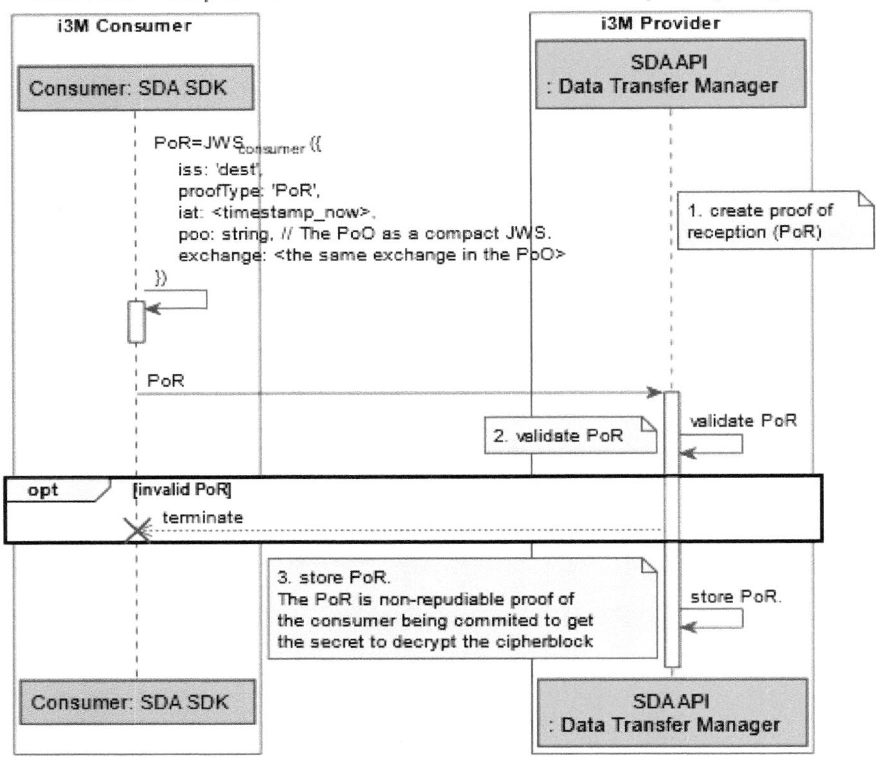

Figure 5.3 NRP — step 2: consumer sends a proof of reception (PoR).

API of the conflict resolver service:

The conflict resolver service implements a HTTP API following the OpenAPI standard. The specification can be consulted in the openapi.json file in the root directory of the conflict resolver service at [61]. For convenience, it can also be visualized online at editor.swagger.io[1] as it is shown in Figure 5.7.

The CRS provides two endpoints: one for checking that the protocol was executed properly, and the other one to initiate a dispute when a consumer claims that he cannot decrypt the cipherblock he has been invoiced for.

[1] https://editor.swagger.io/?url=https://raw.githubusercontent.com/i3-MARKET-V3-Public-Repository/SP3-SCGBSSW-CR-ConflictResolverService/public/spec/openapi.yaml

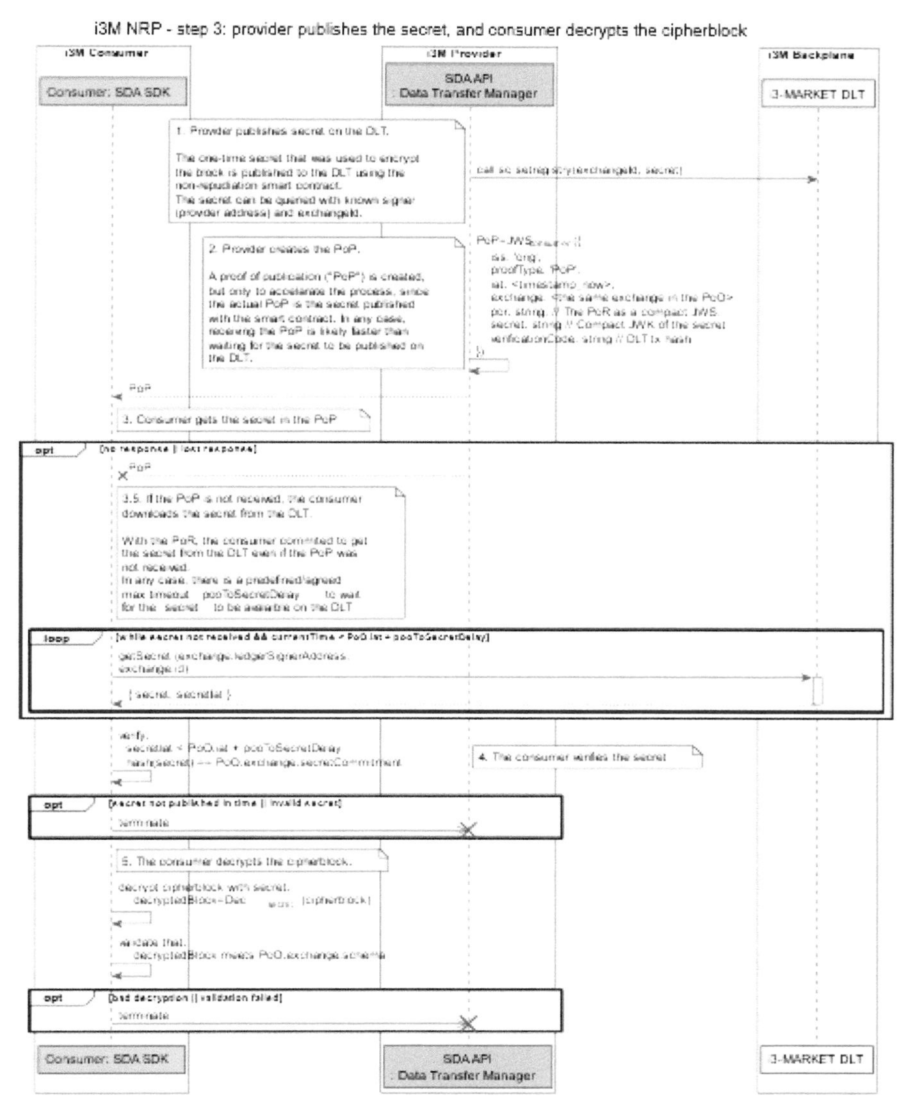

Figure 5.4 NRP — step 3: provider publishes the secret, and consumer decrypts the cipherblock.

The endpoints require JWT bearer authentication. The JWT can be obtained after performing a login with OIDC and presenting valid i3-MARKET credentials.

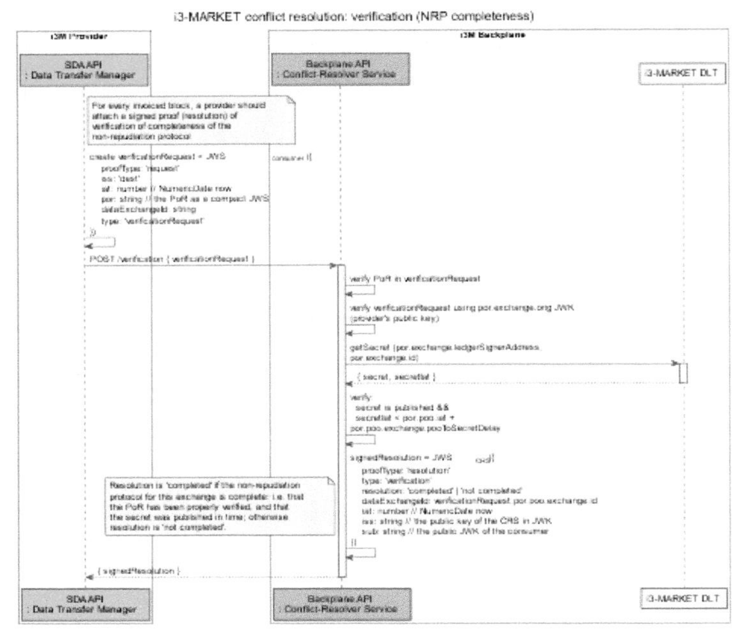

Figure 5.5 Conflict resolution: verification (NRP completeness).

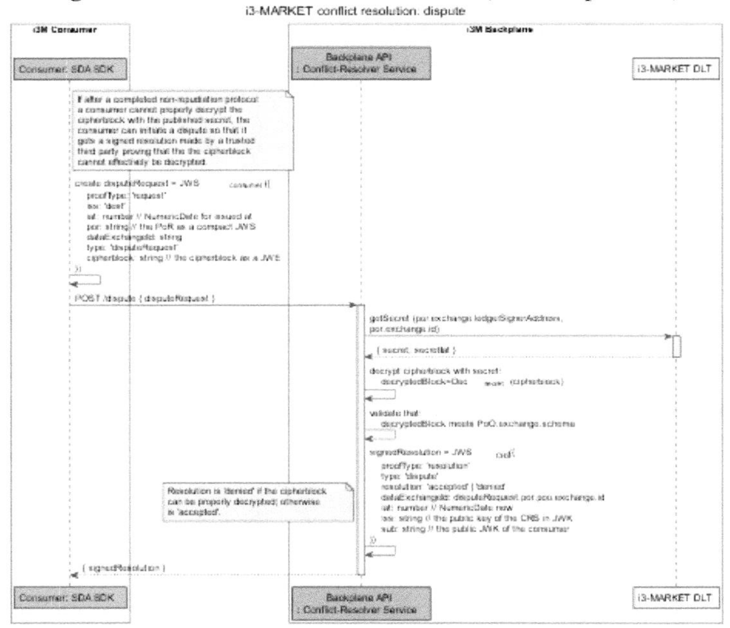

Figure 5.6 Conflict resolution: dispute.

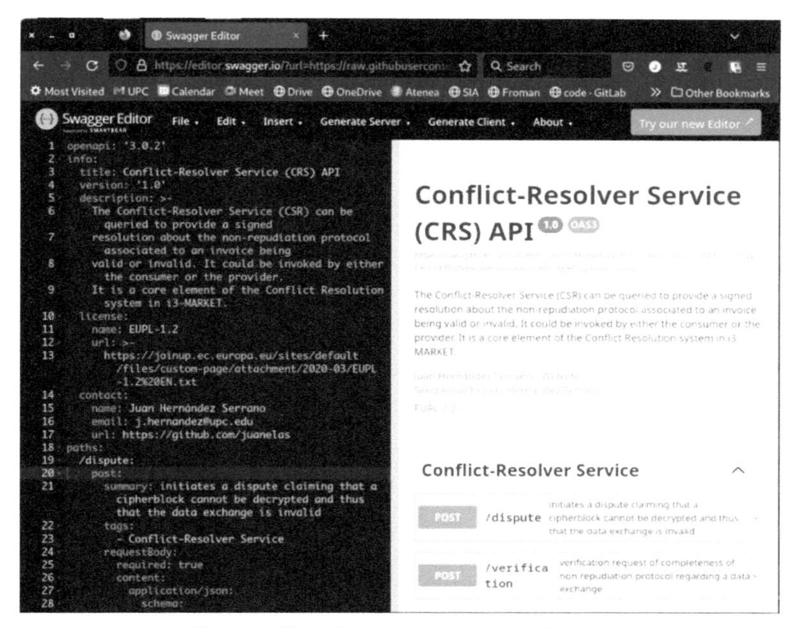

Figure 5.7 CRS API at swagger.editor.io.

- **POST/verification.**

The CRS can be queried to provide a signed resolution about a data exchanged successfully performed or not. It could be invoked by either the consumer or the provider. The provider should query this endpoint and send it along with the invoice to the consumer.

This endpoint can be accessed at POST/verification and requires valid i3-MARKET consumer or provider's credentials.

Input:
A verification request as a compact JSON Web Signature (JWS). For the request to be accepted, it *must* be signed with the same key it was used during the data exchange for this verification.

```
{
  verificationRequest: string // the verification request in compact JWS format
}
```

A verification request is a JWS signed by either the consumer or the provider using the same key he/she used for the data exchange. The verification request payload holds a valid PoR:

```
{
  type: 'verificationRequest'
  proofType: 'request'
  iss: 'orig' | 'dest'
  iat: number // unix timestamp for issued at
  por: string // a compact JWS holding a PoR. The proof MUST be signed with the
same key as either 'orig' or 'dest' of the payload proof.
  dataExchangeId: string // the unique id of this data exchange
}
```

Output:

It returns a signed resolution as a compact JWS with payload:

```
{
  proofType: 'resolution'
  type: 'verification'
  resolution: 'completed' | 'not completed' // whether the data exchange has been
verified to be complete
  dataExchangeId: string // the unique id of this data exchange
  iat: number // unix timestamp stating when it was resolved
  iss: string // the public key of the CRS in JWK
  sub: string // the public key (JWK) of the entity that requested a resolution
}
```

- **POST/dispute.**

Note that the signed resolution obtained from POST/verification does not ensure that the published secret could be used to decrypt the encrypted block of data. If the consumer B is not able to decrypt the cipherblock, he could initiate a dispute on the CRS. The CRS will also provide signed resolution of whether B is right or not.

All this is handled in this endpoint, which can only be queried if in possession of valid i3-MARKET consumer's credentials.

Input:

```
{
  disputeRequest: string // the dispute request in compact JWS format
}
```

A dispute request as a compact JSON Web Signature (JWS). For the request to be accepted, it *must* be signed with the same key it was used during the data exchange for this verification.

The payload of a decoded disputeRequest holds a valid PoR, and the received cipherblock:

```
{
  proofType: 'request'
  type: 'disputeRequest'
  iss: 'dest'
  cipherblock: string // the cipherblock as a JWE string
  iat: number // unix timestamp for issued at
  por: string // a compact JWS holding a PoR. The proof MUST be signed with the
same key as either 'orig' or 'dest' of the payload proof.
  dataExchangeId: string // the unique id of this data exchange
}
```

Output:
It returns a signed resolution as a compact JWS with payload:

```
{
  proofType: 'resolution'
  type: 'dispute'
  resolution: 'accepted' | 'denied' // resolution is 'denied' if the cipherblock
can be properly decrypted; otherwise is 'accepted'
  dataExchangeId: string // the unique id of this data exchange
  iat: number // unix timestamp stating when it was resolved
  iss: string // the public key of the CRS in JWK
  sub: string // the public key (JWK) of the entity that requested a resolution
}
```

5.6 Background Technologies

Both the non-repudiation library and the conflict resolver service need access to a DLT. Access to the DLT is provided by the following technologies:

- Ethers.js [55] is a complete and compact library for interacting with the Ethereum-based DLTs. Along with Web3 is the reference implementation for that purpose.
- Veramo [56] is a JavaScript Framework for Verifiable Data that was designed from the ground up to be flexible and modular, which makes it highly scalable. It can run on several environments: node, mobile, and browser. Its main utility is to make easy the use of DIDs, Verifiable Credentials, and data-centric protocols to bring next-generation features to users.

The smart contracts that regulate the Non-repudiation Protocol have been developed in Solidity [37], an object-oriented, high-level language for implementing smart contracts for Ethereum-like DLTs, and the development environment of choice has been Hardhat.

The non-repudiation library can be instantiated from JavaScript or TypeScript code. It internally uses Panva's JOSE [63] to handle JSON web keys, and Ajv [64] to check and verify JSON schema.

Conflict resolver service HTTP API is developed using Express [65], a minimal and flexible Node.js web application framework that provides a robust set of features for creating robust APIs (among other things).

The conflict resolver service meets the OpenAPI specification [57] with validation of all inputs against the OpenAPI schema using express-openapi-validator [58].

6

Explicit Consent

6.1 Objectives

i3-MARKET's architecture has been designed to allow all the stakeholders – namely providers, consumers, data owners, and marketplace operators – to meet the strictest policies in terms of privacy and data protection, which in fact leads to meet the GDPR requirements with little effort.

Article 4 of the GDPR [26] defines consent as "any freely given, specific, informed and unambiguous indication of the data subject's wishes by which he or she, by a statement or by a clear affirmative action, signifies agreement to the processing of personal data relating to him or her". Data controllers shall be able to demonstrate that they hold the explicit consent of the data subjects to process (Article 7) and/or trade their data. To the best of our knowledge, no technology is enforcing user consent to the point of preventing trading without it.

It is a remarkably innovative feature of the i3-MARKET project that the explicit consent of the data subjects is absolutely required for trading users' data.

6.2 Technical Requirements

The explicit consent subsystem inherits the following technical requirements from the GDPR [26]:

- Trading of sensitive data related to people/entities require their explicit consent.
- The consent can be revoked.
- If a consent is revoked, the data cannot be sold/distributed again.
- The data should be also deleted from already sold datasets.

- The enforcement of the explicit consent should not leak any sensitive data.
- The solution must support non-digitally native data subjects, which delegate consent management to an i3-MARKET provider.

6.3 Solution Design/Blocks

The explicit consent system relies on two main complementary actions: explicit consent and limited data lifetime.

Use-case 1: the data subject is an active i3-MARKET stakeholder:

The explicit consent is a legal agreement between a data provider and the subject of the data. It is out of the scope of i3-MARKET, which is just a technology. However, for every subject involved, a provider should provide an (anonymous) identifier of the consent signed by the subject using an anonymous identity only known to the provider.

As a result, a data offering in i3-MARKET that deals with sensitive data includes a list of signed consents of the data subjects. The smart contract manager (SCM) will verify the consent signatures and status when orchestrating a data sharing agreement. If a consent is not in place or revoked, the SCM prevents the exchange of the affected data.

Obviously, subject can at any time revoke a consent and therefore prevent their data to be sold again. Proving ownership of the consent requires interaction with the SCM using the subject's anonymous identity, which requires the use of the i3M-Wallet.

Note that i3-MARKET does not analyze or check for validity of the actual consent agreements between providers and data subjects. It is the (legal) responsibility of the provider to have the consent in place when legally required. Obviously, the way the consent anonymous ID is created guarantees that the presented consent form was the one registered in i3-MARKET.

For a better understanding on how consents are managed, refer to the detailed diagrams and the SCM explicit-consent endpoints.

Use-case 2: the data subject delegates consent management to an i3-MARKET provider:

The concept of limited data lifetime refers to when a dataset is sold; the consumer accepts the legal obligation to delete it after the agreed lifetime. The lifetime of course must meet the affecting regulation. In i3-MARKET, we are using a 14-day lifetime for the wellbeing pilot.

Meeting the limited data lifetime requirement is out of the scope of i3-MARKET, which just labels datasets with the lifetime. Indeed, its implementation is the responsibility of the data providers and consumers, which should sign legal agreements stating that the dataset should be deleted after the agreed time (lifetime). Note that this does not imply that the consumer cannot access the data again after erasing it; it only means that the consumer will need to re-download them again.

When a data subject revokes consent, the GDPR not only states that her data should not be sold again but also that it should be removed from any sold dataset. Limited data lifetime is absolutely necessary to legally comply with the GDPR. Re-downloading again guarantees that data related to revoked consents "disappears" from any sold dataset.

6.3.1 Diagrams

In the following, we present four sequence diagrams representing the flows for giving consent and revoking it in two use-cases:

1. The data subject is an active i3-MARKET stake holder that can use her own wallet to interact with the i3-MARKET Backplane.
2. The data subject delegates consent management to the data provider, which will therefore interact with the i3-MARKET Backplane on behalf of the subject.

Use-case 2 meets the last technical requirement, introduced by the i3-MARKET wellbeing pilot, which states that data subjects may not be digitally natives or may not be interested in being an active i3-MARKET stakeholder.

The diagrams in Figures 6.1–6.4 are self-explanatory, but consider analyzing the SCM endpoints for the explicit consent for a better understanding of the flow.

Explicit consent:

- **Giving consent:**

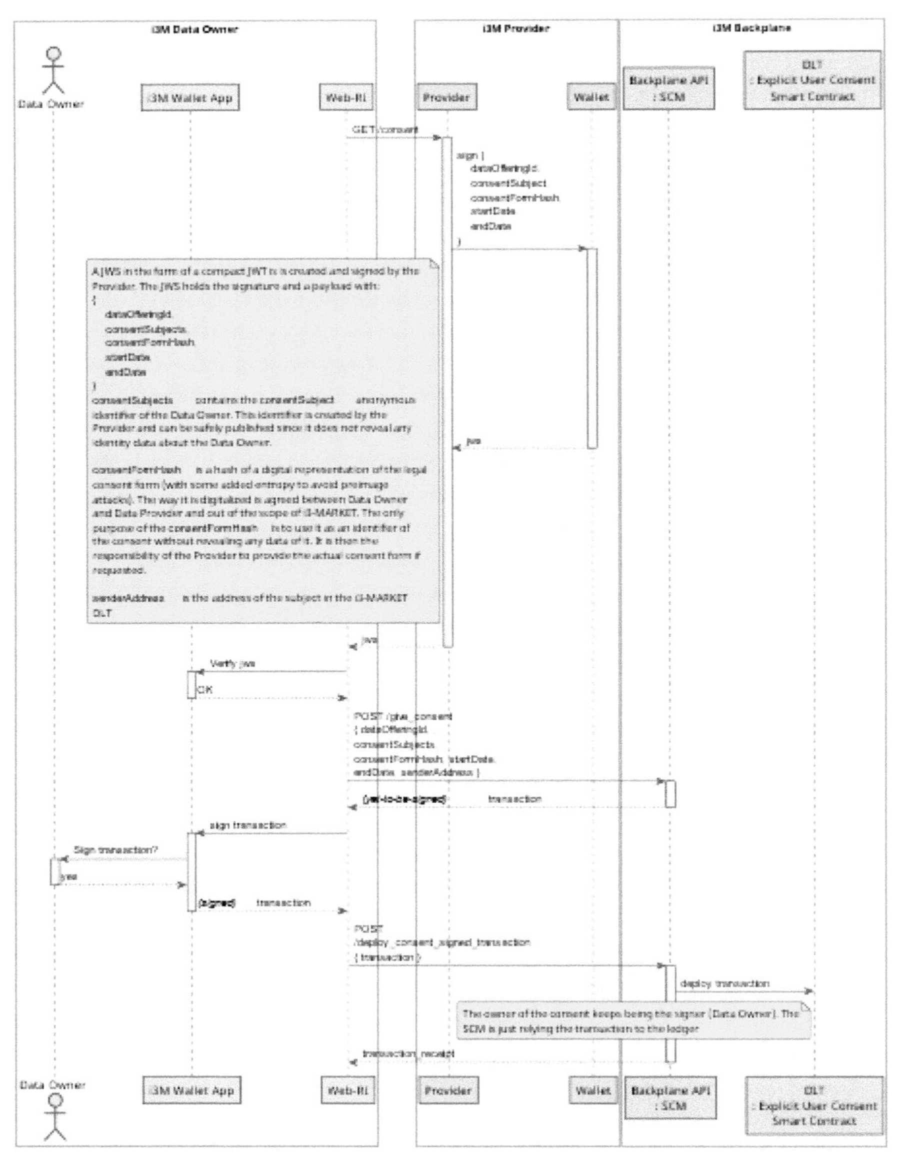

Figure 6.1 Use-case 1: giving explicit consent.

- **Revoking consent:**

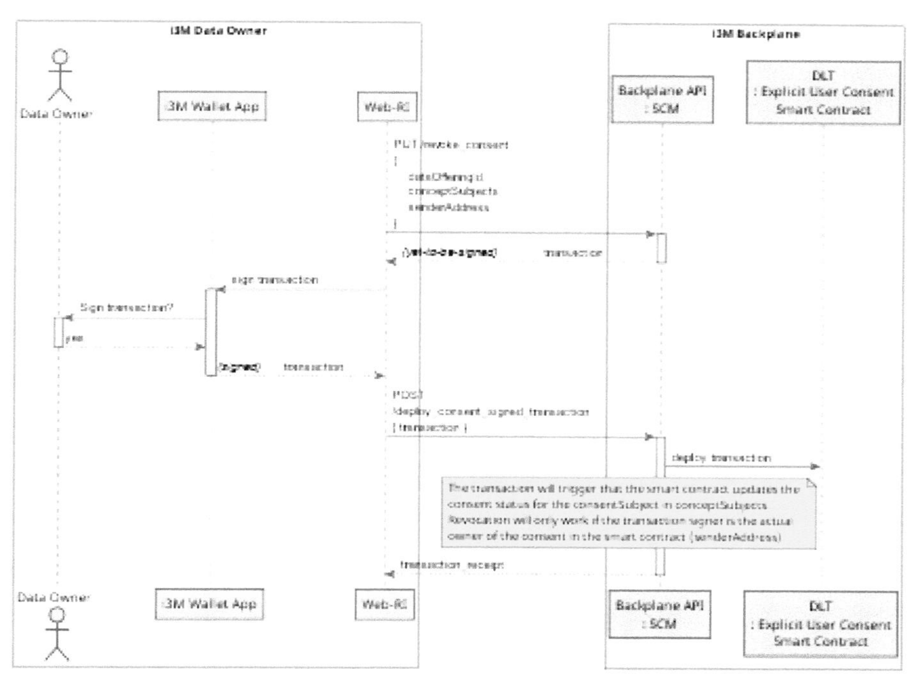

Figure 6.2 Use-case 1: revoking consent.

Limited data lifetime:

- **Giving consent:**

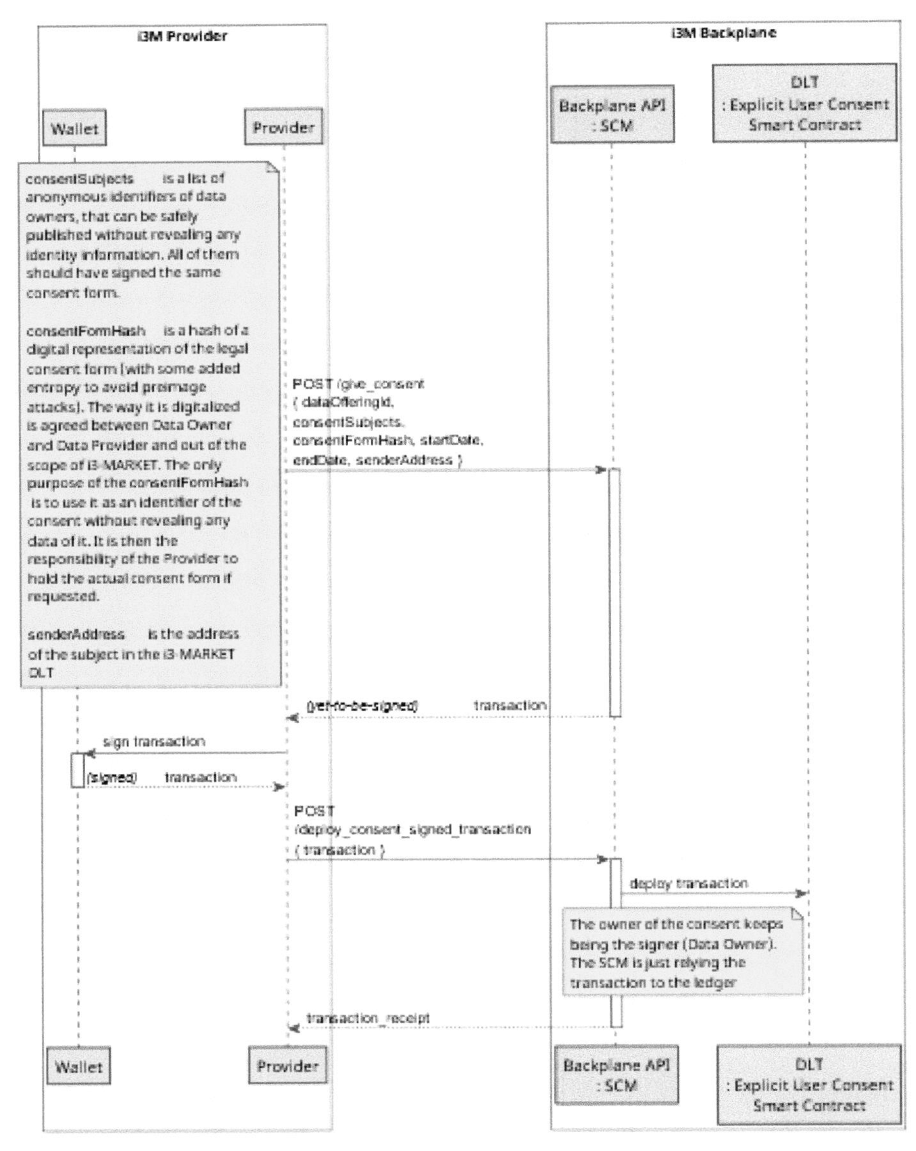

Figure 6.3 Use-case 2: giving explicit consent.

- **Revoking consent:**

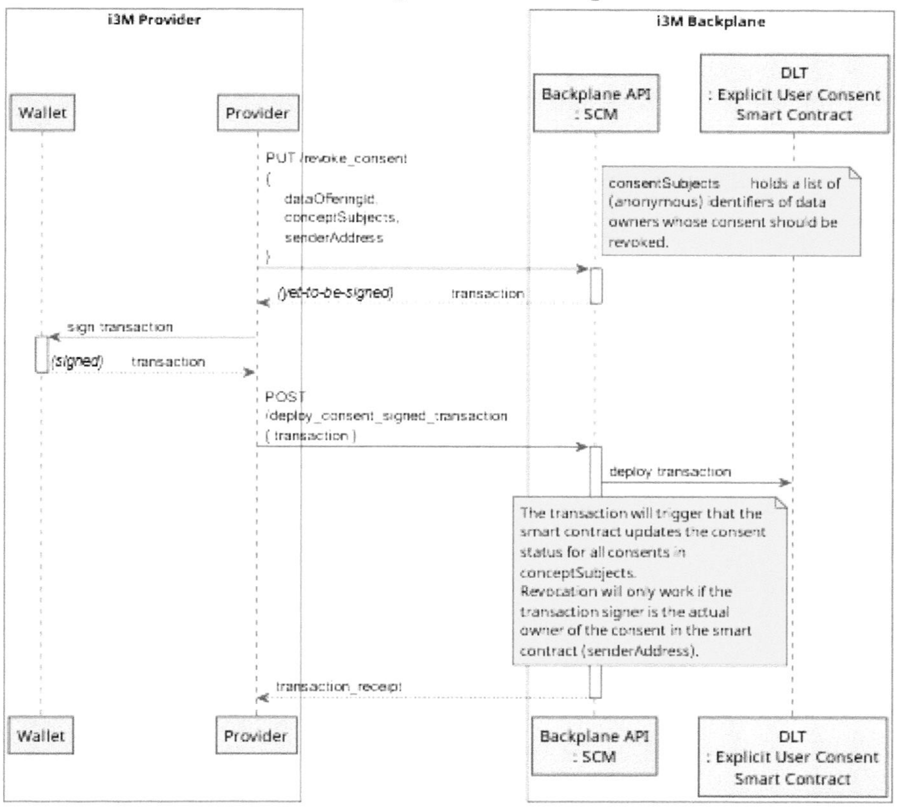

Figure 6.4 Use-case 2: revoking consent.

6.4 Background Technologies

The explicit user consent subsystem has no special selected technologies since its development is actually split into other subsystems:

- Consent giving and revoking is implemented in the smart contract manager, which is described in more detail in Chapter 9.
- Consent checking before exchanging data related to a data subject is implemented between the smart contract manager and the secure data access SDK.

7

Smart Contract Manager

7.1 Objectives

The smart contract manager (SCM) provides a gateway to access the smart contracts and is used by other subsystems to integrate their functionalities (conflict resolution, pricing manager, explicit user consent, and secure data exchanges).

Smart contract manager facilitates the creation of agreement objects using the data sharing agreement (DSA) smart contract. The DSA solidity contract is based on a legal agreement for data sharing, considering the existing legal framework (e.g., GDPR [26]). The agreement objects are used to enforce agreed-upon obligations from the provider and consumer sides.

The smart contract manager development has been made publicly available in the i3-MARKET GitHub repository and the smart contracts the subsystem uses at [66]. The Table 7.1 summarizes the Smart Contract Manager user stories.

7.2 Technical Requirements

Table 7.1 Smart contract manager – user stories.

Name	*Description*	*Labels*
SCM	Within i3-MARKET, DSA objects need to be stored on the blockchain in order to automatically enforce certain clauses of the legal data trading agreement. Additionally, automatic conflict resolution of certain types of violations has to be supported. The smart contracts of the SCM need to combine legal certainty with automated enforcement, built-in conflict resolution mechanisms, and guaranteed access to remedy. The SCM evaluates a signed resolution, issued by the conflict-resolver service, which relies on the execution of the Non-repudiation Protocol. Depending on the type of resolution, the state of the agreement is automatically updated.	User story

Table 7.1 *Continued.*

Name	*Description*	*Labels*
	Explicit data owner consent: In case of personal data, legal consent of data owners is required. When the consent is given, the SCM stores a list of explicit consents for a specific offering. The consent can be revoked anytime, and before an agreement is created, the consent status is verified. As long as the data to be shared is personal data, agreements can be created just when the consent was given by the data owner. Pricing: The price and the fee of the data are stored in the agreement. The fee is requested from the pricing manager, based on the price in the data offering.	

7.3 Solution Design/Blocks

The smart contract manager extracts the contractual parameters from the data offering description and returns a template with possible contractual parameters (to be displayed in the marketplace), as shown in Figure 7.1. After a data purchase request is sent, with a potential proposal of new parameters by the consumer, the provider and consumer must sign the agreement and store it in the wallet. As soon as both received the signed data sharing agreement and saved it in the wallet, the provider can create and store the agreement on the blockchain. The smart contract manager invokes the data sharing agreement smart contract and creates an agreement with the proposed contractual parameters. The agreement object is put on the ledger and automatically enforced by the corresponding smart contract (Figure 7.2).

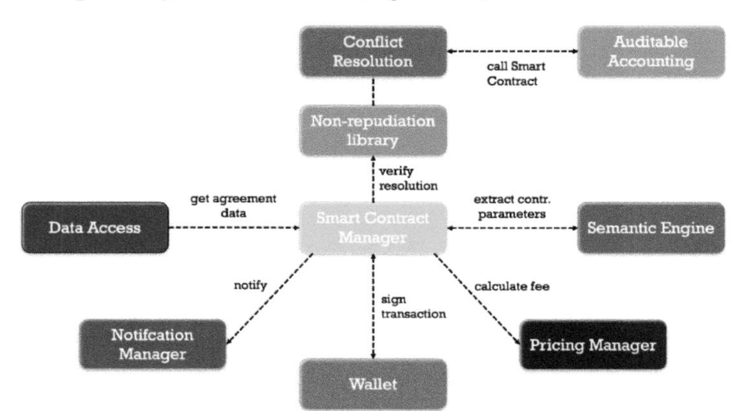

Figure 7.1 Context view of the smart contract manager.

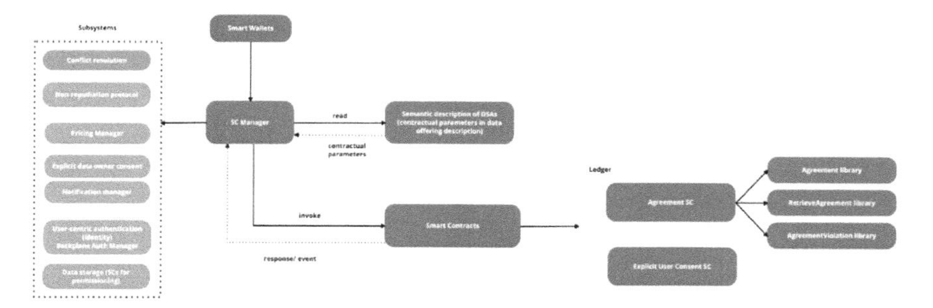

Figure 7.2 Component diagram of the smart contract manager subsystem.

The smart contract manager is interconnected with the following i3-MARKET subsystems, as it is shown in Figures 7.3–7.5.

- **Semantic engine:** To retrieve the parameters and details about the data offering descriptions to compile information for the contract agreements.
- **Conflict resolution:** In order to check whether a violation to the contract occurred, the conflict resolution is invoked. The conflict resolution will prevent any two peers of a data exchange, namely provider and consumer to deny that a given data-block exchange happened or to assert that a data-block exchange that did not happen, happened. The conflict-resolver service issues verifiable signed resolutions regarding the execution of the i3-MARKET Non-Repudiation Protocol. The SCM evaluates the signed resolution and, depending on the type of resolution, automatically changes the state of the agreement in case of a violation, as well as suggests penalties for one of the peers.
- **Non-repudiation Protocol:** The Non-repudiation Protocol aims at preventing parties in a data exchange from falsely denying having taken part in that exchange.
- **Explicit data-owner consent:** To ensure an explicit consent of the data owners every time their personal data is traded, the explicit data owner consent component is triggered.
- **Pricing manager:** The SCM requests the fee of the data based on the price registered in the data offering by invoking the pricing manager to calculate the corresponding fee and includes it in the contractual template.
- **User-centric authentication:** To ensure that only authorized participants (with the corresponding role) are able to trigger functionality

provided by the data sharing agreement smart contract (via the smart contract manager), user-centric authentication is used (part of the Backplane).

- **i3M-Wallet:** The raw transactions created in the SCM have to be signed with an i3M-Wallet (either the Wallet Desktop App or the server wallet) in order to deploy them.

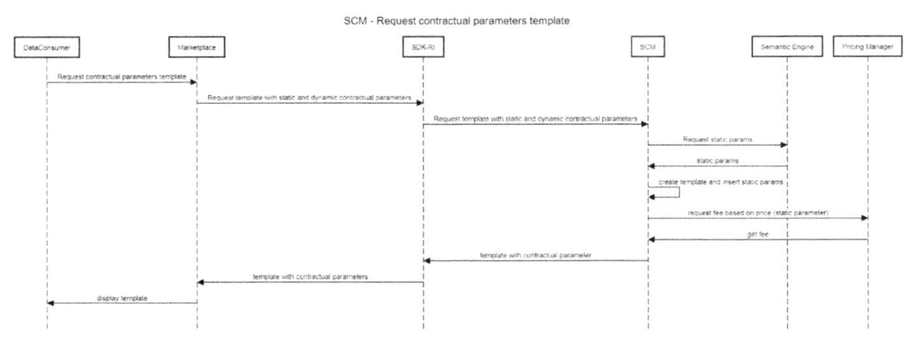

Figure 7.3 Sequence diagram – retrieve contractual parameters template.

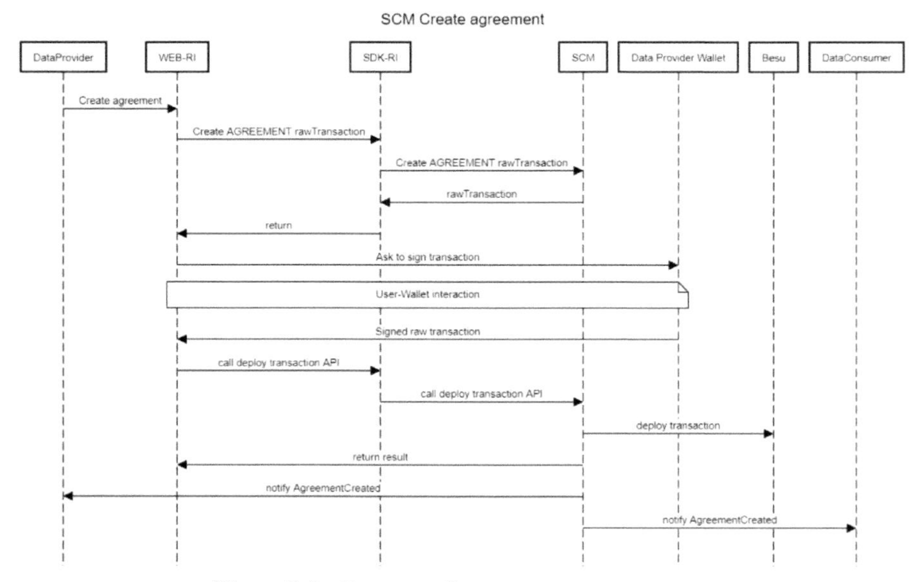

Figure 7.4 Sequence diagram – create agreement.

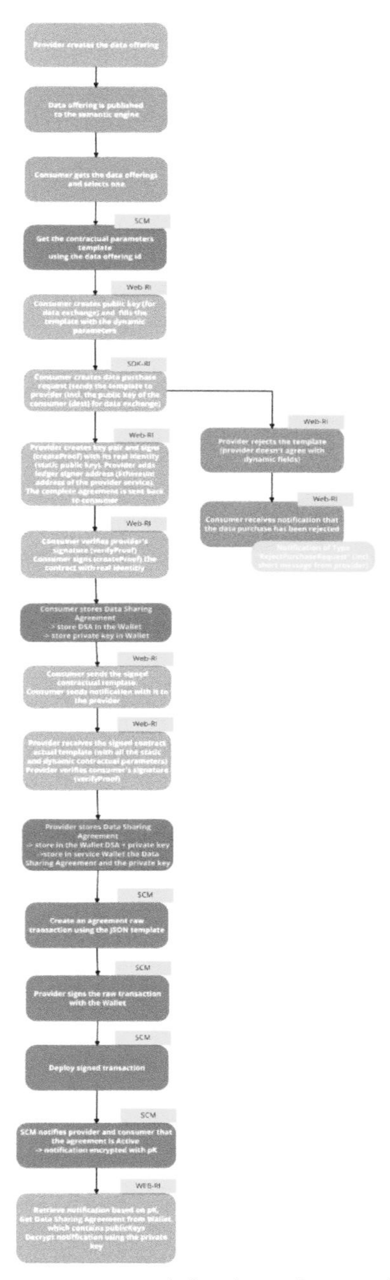

Figure 7.5 Data sharing agreement negotiation, key pair generation, storage in wallet, and agreement creation on blockchain.

7.4 Diagrams

The smart contract manager extracts the static contractual parameters from the data offering description using the semantic data model. The interactions are shown in Figure 7.6. The dynamic parameters, such as the consumer DID, start date, and end date of the agreement, are filled when a data purchase request is created by the consumer.

Before storing an agreement on the blockchain using the smart contract manager, the provider and the consumer should generate their public–private keys (using the non-repudiation library) and they should each sign the contract. After they filled in their public keys and the contract is signed, they should store the generated key pairs and data sharing agreement in their wallets as shown in Figure 7.7.

As soon as the negotiation between the provider and consumer is over and they agree on specific contractual parameters, as well as store the final data sharing agreement and the key pairs in their wallets, the provider can create the agreement on the blockchain using the smart contract manager.

Firstly, a raw transaction is created using the data sharing agreement, which was saved in the wallet. The successful response of creating an agreement request is a raw transaction object. This raw transaction has to be signed with the wallet using the provider's DID. After the signed transaction is obtained from the wallet, it has to be deployed. The response of the Smart Contract Manager should be a transaction object with information about

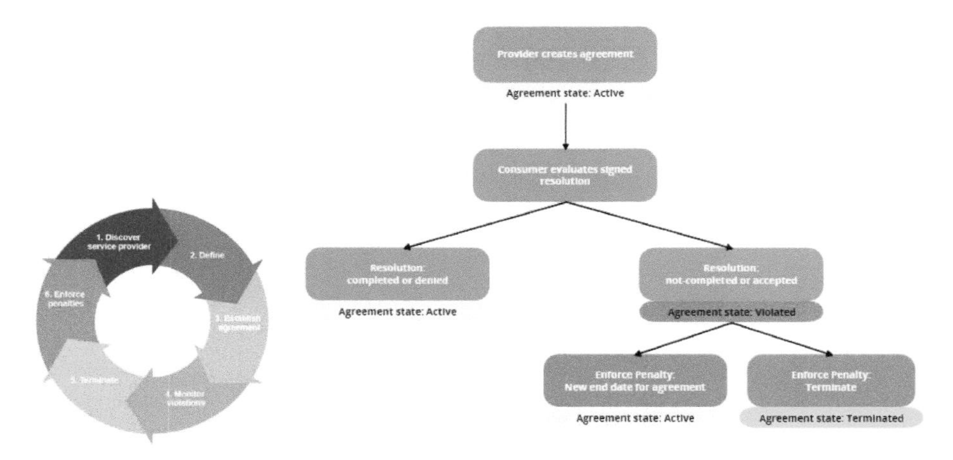

Figure 7.6 Sequence diagram – check agreements by offering ID.

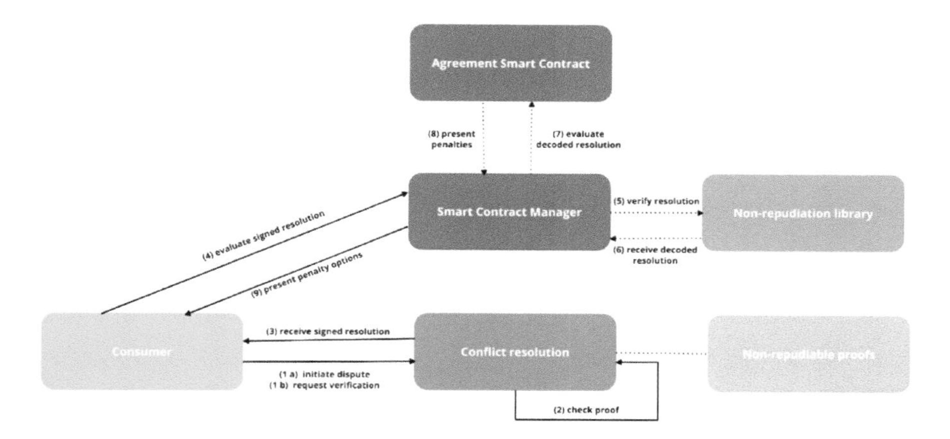

Figure 7.7 Conflict resolution.

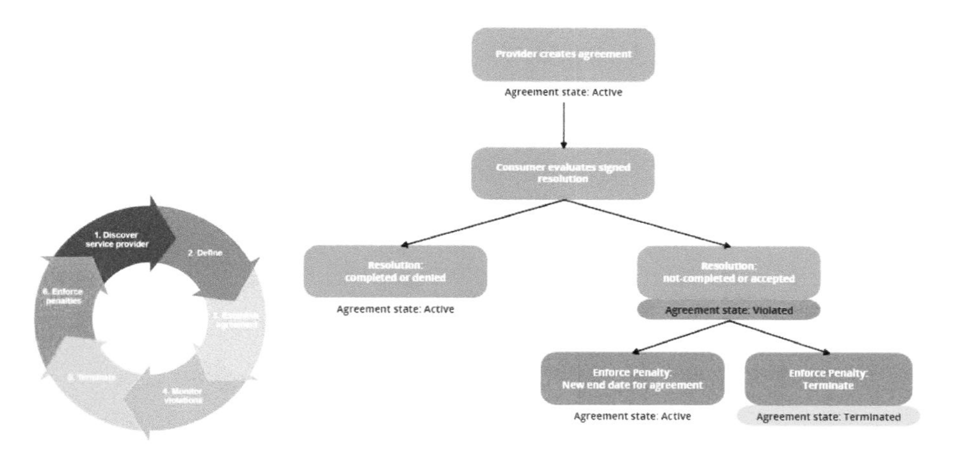

Figure 7.8 Agreement lifecycle and states.

the transaction in Figure 7.8. If the confirmation is 1, the transaction was successfully deployed, and the agreement is stored on the blockchain.

After that, the provider and consumer receive a notification that the agreement is active, which means it was created and stored on the blockchain. This notification will be encrypted and contains the agreement id. The notifications should be retrieved from the notification manager based on the provider/consumer public key and decrypted using the corresponding private key. After they receive this notification, the provider should *post* the data

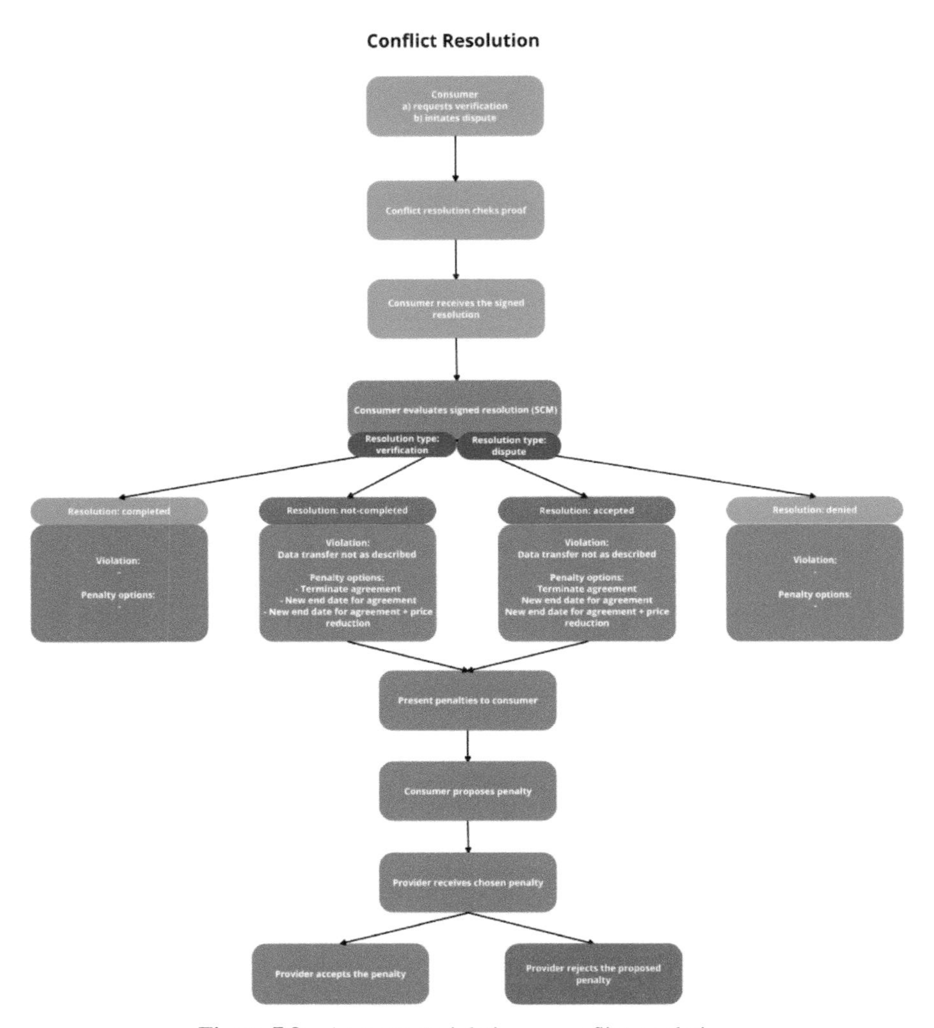

Figure 7.9 Agreement violation – conflict resolution.

exchange agreement, the agreement id, and the private key to data access and then the consumer can start the transfer – see Figure 7.9.

Agreement violation – conflict resolution:

After the data transfer is finished, a consumer can request a verification or initiate a dispute using the conflict resolution. The proof of the completeness of the data exchange will be checked and the consumer receives the signed resolution based on that proof.

The smart contract manager evaluates the signed resolution. Within this evaluation, the resolution is decoded and depending on the resolution, the agreement's state can change from active to violated.

The transfer was unsuccessful when the resolution is:

- not completed (in case of a verification) – the decryption key was not published;
- accepted (in case of a dispute) – the cypher block cannot be properly decrypted.

If the transfer was not successful, the agreement is violated. When the agreement is violated, the consumer receives a list of penalties.

These penalties could be:

- new end date for agreement;
- new end date for agreement and a price reduction;
- termination of agreement.

The consumer should propose one of these penalties to the provider. The provider will receive a notification with the chosen penalty and if he agrees to the penalty, he should enforce on the blockchain. By enforcing the new penalty, the agreement state changes from violated to active or terminated (in case the penalty termination is chosen).

7.5 Interfaces

The smart contract manager API is the interface via which the clients gain access to the smart contract parameters.

The endpoints documented below were grouped by modules.

Agreement:

`GET /template/{offering_id}`

Request template with static and dynamic parameters

offering_id (required)

Example data

Content-Type: application/json


```
{
  "dataOfferingDescription": {
    "dataOfferingId": "63662ebdb7d5dd78b7159566",
    "version": 0,
```

```
    "title": "Oil Supply Unit",
    "category": "manufacturing",
    "active": true
  },
  "parties": {
    "providerDid":
"did:ethr:i3m:0x0243cc9dbc7157ee12ce1898ac0c49b366822f32d57bc108e127f45b6c4
3a57e90",
    "consumerDid": "string"
  },
  "purpose": "Oil supply Unit measurements",
  "duration": {
    "creationDate": 0,
    "startDate": 0,
    "endDate": 0
  },
  "intendedUse": {
    "processData": true,
    "shareDataWithThirdParty": false,
    "editData": true
  },
  "licenseGrant": {
    "transferable": false,
    "exclusiveness": true,
    "paidUp": true,
    "revocable": true,
    "processing": true,
    "modifying": true,
    "analyzing": true,
    "storingData": true,
    "storingCopy": true,
    "reproducing": true,
    "distributing": false,
    "loaning": false,
    "selling": false,
    "renting": false,
    "furtherLicensing": false,
    "leasing": false
  },
  "dataStream": false,
  "personalData": false,
  "pricingModel": {
    "paymentType": "one-time purchase",
    "pricingModelName": "string",
    "basicPrice": 125.68,
    "currency": "$",
    "fee": 6.28,
    "hasPaymentOnSubscription": {
      "paymentOnSubscriptionName": "",
      "paymentType": "",
      "timeDuration": "",
      "description": "",
      "repeat": "",
      "hasSubscriptionPrice": 0
    },
    "hasFreePrice": {
      "hasPriceFree": false
    }
  },
  "dataExchangeAgreement": {
    "orig": "string",
```

```
    "dest": "string",
    "encAlg": "A128GCM",
    "signingAlg": "ES256",
    "hashAlg": "SHA-256",
    "ledgerContractAddress": "0x8d407a1722633bdd1dcf221474be7a44c05d7c2f",
    "ledgerSignerAddress":
"0x02897978ebd80646bc469cba19d79d8655cd862cb9fd2484141d66103260cc540d",
    "pooToPorDelay": 100000,
    "pooToPopDelay": 30000,
    "pooToSecretDelay": 180000
  },
  "signatures": {
    "providerSignature": "string",
    "consumerSignature": "string"
  }
}
```


Returns the template with static and dynamic contractual parameters

`POST /sdk-ri/contract/create-data-purchase`

Create data purchase request (not part of the Backplane) – sends notification to provider with the static and dynamic parameters filled in by the consumer

`POST /create_agreement_raw_transaction/{sender_address}`

Create agreement raw transaction (createAgreement)
sender_address (required)
Request body
body template (required)


```
{
  "dataOfferingDescription": {
    "dataOfferingId": "63662ebdb7d5dd78b7159566",
    "version": 0,
    "title": "Oil Supply Unit",
    "category": "manufacturing",
    "active": true
  },
  "parties": {
    "providerDid":
"did:ethr:i3m:0x0243cc9dbc7157ee12ce1898ac0c49b366822f32d57bc108e127f45b6c4
3a57e90",
    "consumerDid":
"did:ethr:i3m:0x03878572e4476a6b7b0223d07f53159ef923c874084ea56760fd130d80c
51409ad"
  },
  "purpose": "P&ID diagram of the Lube Oil supply Unit",
  "duration": {
    "creationDate": 1678997655,
    "startDate": 1786678869,
    "endDate": 1886678869
  },
  "intendedUse": {
    "processData": true,
    "shareDataWithThirdParty": false,
    "editData": true
  },
```

```
  "licenseGrant": {
    "transferable": false,
    "exclusiveness": false,
    "paidUp": true,
    "revocable": true,
    "processing": true,
    "modifying": true,
    "analyzing": true,
    "storingData": true,
    "storingCopy": true,
    "reproducing": true,
    "distributing": false,
    "loaning": false,
    "selling": false,
    "renting": false,
    "furtherLicensing": false,
    "leasing": false
  },
  "dataStream": false,
  "personalData": false,
  "pricingModel": {
    "paymentType": "one-time purchase",
    "pricingModelName": "string",
    "basicPrice": 125.68,
    "currency": "$",
    "fee": 6.28,
    "hasPaymentOnSubscription": {
      "paymentOnSubscriptionName": "string",
      "paymentType": "string",
      "timeDuration": "string",
      "description": "string",
      "repeat": "string",
      "hasSubscriptionPrice": 0
    },
    "hasFreePrice": {
      "hasPriceFree": false
    }
  },
  "dataExchangeAgreement": {
    "orig":                                    "{\"kty\":\"EC\",\"crv\":\"P-
256\",\"x\":\"4sxPPpsZomxPmPwDAsqSp94QpZ3iXP8xX4VxWCSCfms\",\"y\":\"8YI_bvV
rKPW63bGAsHgRvwXE6uj3TlnHwoQi9XaEBBE\",\"alg\":\"ES256\"}",
    "dest":                                    "{\"kty\":\"EC\",\"crv\":\"P-
256\",\"x\":\"6MGDu3EsCdEJZVV2KFhnF2lxCRI5yNpf4vWQrCIMk5M\",\"y\":\"0OZbKAd
ooCqrQcPB3Bfqy0g-Y5SmnTyovFoFY35F00N\",\"alg\":\"ES256\"}",
    "encAlg": "A256GCM",
    "signingAlg": "ES256",
    "hashAlg": "SHA-256",
    "ledgerContractAddress": "0x7B7C7c0c8952d1BDB7E4D90B1B7b7C48c13355D1",
    "ledgerSignerAddress": "0x17bd12C2134AfC1f6E9302a532eFE30C19B9E903",
    "pooToPorDelay": 10000,
    "pooToPopDelay": 20000,
    "pooToSecretDelay": 150000
  },
  "signatures": {
     "providerSignature":
"eyJhbGciOiJQUzM4NCIsImtpZCI6ImJpbGJvLmJhZ2dpbnNAaG9iYml0b24uZXhhbXBsZSJ9.S
XTigJlzIGEgZGFuZ2Vyb3VzIGJ1c2luZXNzLCBGcm9kbywgZ29pbmcgb3V0IHlvdXIgZG9vci4g
WW91IHN0ZXAgb250byB0aGUgcm9hZCwgYW5kIGlmIHlvdSBkb24ndCBrZWVwIHlvdXIgZmVldCw
gdGhlcmXigJlzIG5vIGtub3dpbmcgd2hlcmUgeW91IG1pZ2h0IGJlIHN3ZXB0IG9mZiB0by4.cu
22eBqkYDKgIlTpzDXGvaFfz6WGoz7fUDcfT0kkOy42miAh2qyBzk1xEsnk2IpN6tPid6VrklHkq
```

```
sGqDqHCdP6O8TTB5dDDItllVo6_1pcbUrhiUSMxbbXUvdvWXzg-
UD8biiReQFlfz28zGWVsdiNAUf8ZnyPEgVFn442ZdNqiVJRmBqrYRXe8P_ijQ7p8Vdz0TTrxUeT
3lm8d9shnr2lfJT8ImUjvAA2Xez2Mlp8cBE5awDzT0qI0n6uiPlaCN_2_jLAeQTlqRHtfa64QQS
UmFAAjVKPbByi7xho0uTOcbH510a6GYmJUAfmWjwZ6oD4ifKo8DYM-X72Eaw",
    "consumerSignature":
"eyJhbGciOiJQUzM4NCIsImtpZCI6ImJpbGJvLmJhZ2dpbnNAaG9iYml0b24uZXhhbXBsZSJ9.S
XTigJlzIGEgZGFuZ2Vyb3VzIGJ1c2luZXNzLCBGcm9kbywgZ29pbmcgb3V0IHlvdXIgZG9vci4g
WW91IHN0ZXAgb250byB0aGUgcm9hZCwgYW5kIGlmIHlvdSBkb24ndCBrZWVwIHlvdXIgZmVldCw
gdGhlcmXigJlzIG5vIGtub3dpbmcgd2hlcmUgeW91IG1pZ2h0IGJlIHN3ZXB0IG9mZiB0by4.cu
22eBqkYDKgIlTpzDXGvaFfz6WGoz7fUDcfT0kkOy42miAh2qyBzk1xEsnk2IpN6tPid6VrklHkq
sGqDqHCdP6O8TTB5dDDItllVo6_1pcbUrhiUSMxbbXUvdvWXzg-
UD8biiReQFlfz28zGWVsdiNAUf8ZnyPEgVFn442ZdNqiVJRmBqrYRXe8P_ijQ7p8Vdz0TTrxUeT
3lm8d9shnr2lfJT8ImUjvAA2Xez2Mlp8cBE5awDzT0qI0n6uiPlaCN_2_jLAeQTlqRHtfa64QQS
UmFAAjVKPbByi7xho0uTOcbH510a6GYmJUAfmWjwZ6oD4ifKo8DYM-X72Eaw"
  }
}
```


Return type
raw_transaction

Example data
Content-Type: application/json


```
{
  "nonce": 46,
  "to": "0x4d722c3a1Cec530671063710349 5dDd9DFAda905",
  "from": "0xc6b8cf76bd7078e56c6ce8c357dd91caeea70170",
  "gasLimit": 12500000,
  "gasPrice": 1000,
  "chainId": 1337,
  "data":
"0x667a8beb0000000000000000000000000000000000000000000000000000000000000000
00000000000000000000000000000000000000000000000000000000000000a000000000000
000000000000000000000000000000000000000000000000000000000000000000000000000
000000000000000000000000000000000000000000000000000000000000000000000000000
000000000000000000000000000000000000000000000000000000000000000000000000000
00000000000000000006737472696e670000000000000000000000000000000000000000000
000000000"
}
```


Returns a raw transaction for the create agreement operation

```
POST /deploy_signed_transaction
```

Deploy signed transaction and send encrypted notification based on the event emitted by the DataSharingAgreement smart contract
Request body
body signed_transaction (required)

Example data
Content-Type: application/json


```
{
  "signedTransaction":
"0xf90f2a2e8203e883bebc20944d722c3a1cec5306710637103495ddd9dfada90580b90ec4
ee4b2db50000000000000000000000000000000000000000000000000000000000003a0000
00000000000000000000000000000000000000000000000079223a224543222c22637276223
a22502d323536222c2278223a22364d4744753345734364454a5a5656324b46686e46326c78
43524935794e7066347657517243494d6b354d222c2279223a22304f5a624b41646f6f43717
25163504233426671793067 2d5935536d6e54796f76466f465933354630304e222c22616c67
223a224553323536227d0000000000000000000000000000000000000000000000000000000
000000000000000000000000000000000000000000000040363133353064633366666643730
32626239373933366338393638636 4694e415566385a6e7950456756466e3434325a644e716
9564a526d427172595258653850 5f5153556d4641416a564b5062427969377868 6f3075544f
000000000000000000000000000000000000000000000000000000000000000000000000000
000000000000000000000000000000000000000000000000000000000000000000000000000
000000000000000000000000000000000000000000000000000000000000000000000673747
2696e670000000000000000000000000000000000000000000000000000820a95a05263ad3d
490c6ab7baf8d755814ece3390de10e7df0cfc1ef3ae58361f949429a056fec9bcb23e8c1a1
cfd7d30f1c4959e63c1863ef1261b5941a9a22d779e855d"
}
```


Return type
transaction_object
Example data


```
{
  "transactionHash":
"0x833013a9428427016fc4b3cd1f05e9b42b289f4f98cd5bccfb91f4ae45fd630d",
  "transactionIndex": 0,
  "blockHash":
"0x1fd6a7de60041d0ec9c4735b9ecd8b022e8cbb154bc4f153cf9c517bc8f7e381",
  "blockNumber": 661175,
  "contractAddress": null,
  "cumulativeGasUsed": 1672030,
  "to": "0x4d722c3a1Cec5306710637103495dDd9DFAda905",
  "from": "0xC6b8cf76BD7078e56C6CE8C357dD91caeEa70170",
  "gasUsed": 1672030,
  "logsBloom":
"0x000000000000000000000000000000000000000000000000000000000000000000000000
000000000000000000000000000000000000000000000000000000000000000000000000000
000000000024000000000000000000000000000000000000000000000000800000000000000
000000000000000000000000000000000000000000000000000000000000000000000000000
000000000000000001000000000000000000000000000000000000000000000000080000000
000000000000000080000000000000000000000000000000000000000000000000000000000
00000000000000000000000000000000000000000000000000000000000000000",

"logs": [
    {
      "transactionIndex": 0,
      "blockNumber": 661175,
      "transactionHash":
"0x833013a9428427016fc4b3cd1f05e9b42b289f4f98cd5bccfb91f4ae45fd630d",
      "address": "0x4d722c3a1Cec5306710637103495dDd9DFAda905",
      "topics": [

"0x40f080228d46fb72660eddafe315e4a5b47df236dc33b76fcd122bcbea89b01d"
      ],
      "data":
"0x000000000000000000000000000000000000000000000000000000000000006000000000
```

```
0000000000000000000000000000000000000000000000000120000000000000000000
000000000000000000000000000000000000000000f000000000000000000000000000
000000000000000000000000000008c7b226b7479223a224543222c22637276223a22502
d323536222c2278223a223473785050707 35a6f6d78506d50774441737153703934517 05a33
695850387858345678574353 43666d73222c2279223a223859495f627656724b50573633624
74173486752767758453675 6a33546c6e48776f51693958614542424 5222c22616c67223a22
4553323536227d0000000000000000000000000000000000000000000000000000000
000000000000000000000000000000000000008c7b226b7479223a224543222c22637276
223a22502d323536222c2278223a22364d4744753345734364454a5a5656324b46686e46326
c7843524935794e70663476575172434 94d6b354d222c2279223a22304f5a624b41646f6f43
71725163504233426671793067 2d5935536d6e54796f76466f465933354630304e222c22616
c67223a224553323536227d00000000000000000000000000000000000000",
      "logIndex": 0,
      "blockHash":
"0x1fd6a7de60041d0ec9c4735b9ecd8b022e8cbb154bc4f153cf9c517bc8f7e381"
    }
  ],
  "confirmations": 1,
  "status": 1
}
```


Returns transaction receipt with confirmation 1

```
GET /get_agreement/{agreement_id}
```

Retrieve an agreement by agreement id
Path parameters
agreement_id (required)
Example data
Content-Type: application/json


```
  {
  "agreementId": 15,
  "providerPublicKey":                          "{\"kty\":\"EC\",\"crv\":\"P-
256\",\"x\":\"4sxPPpsZomxPmPwDAsqSp94QpZ3iXP8xX4VxWCSCfms\",\"y\":\"8YI_bvV
rKPW63bGAsHgRvwXE6uj3TlnHwoQi9XaEBBE\",\"alg\":\"ES256\"}",
  "consumerPublicKey":                          "{\"kty\":\"EC\",\"crv\":\"P-
256\",\"x\":\"6MGDu3EsCdEJZVV2KFhnF2lxCRI5yNpf4vWQrCIMk5M\",\"y\":\"0OZbKAd
ooCqrQcPB3Bfqy0g-Y5SmnTyovFoFY35F00N\",\"alg\":\"ES256\"}",
  "dataExchangeAgreementHash":
"61350dc3ffd702bb97936c8968d9fc19629a427157d6254bea5d415616edf07e",
  "dataOffering": {
    "dataOfferingId": "63662ebdb7d5dd78b7159566",
    "dataOfferingVersion": 0,
    "dataOfferingTitle": "Oil Supply Unit"
  },
  "purpose": "P&ID diagram of the Lube Oil supply Unit",
  "state": 0,
  "agreementDates": [
    1671753600,
    1786678869,
    1886678869
  ],
  "intendedUse": {
    "processData": true,
    "shareDataWithThirdParty": false,
    "editData": true
  },
```

```
  "licenseGrant": {
    "transferable": false,
    "exclusiveness": true,
    "paidUp": true,
    "revocable": true,
    "processing": true,
    "modifying": true,
    "analyzing": true,
    "storingData": true,
    "storingCopy": true,
    "reproducing": true,
    "distributing": false,
    "loaning": false,
    "selling": false,
    "renting": false,
    "furtherLicensing": false,
    "leasing": false
  },
  "dataStream": false,
  "personalData": false,
  "pricingModel": {
    "paymentType": "one-time purchase",
    "price": 125.68,
    "currency": "$",
    "fee": 6.28,
    "paymentOnSubscription": {
      "timeDuration": "string",
      "repeat": "string"
    },
    "isFree": false
  },
  "violation": {
    "violationType": 0
  },
  "signatures": {
    "providerSignature":
"eyJhbGciOiJQUzM4NCIsImtpZCI6ImJpbGJvLmJhZ2dpbnNAaG9iYml0b24uZXhhbXBsZSJ9.S
XTigJlzIGEgZGFuZ2Vyb3VzIGJ1c2luZXNzLCBGcm9kbywgZ29pbmcgb3V0IHlvdXIgZG9vci4g
WW91IHN0ZXAgb250byB0aGUgcm9hZCwgYW5kIGlmIHlvdSBkb24ndCBrZWVwIHlvdXIgZmVldCw
gdGhlcmXigJlzIG5vIGtub3dpbmcgd2hlcmUgeW91IG1pZ2h0IGJlIHN3ZXB0IG9mZiB0by4.cu
22eBqkYDKgIlTpzDXGvaFfz6WGoz7fUDcfT0kkOy42miAh2qyBzk1xEsnk2IpN6tPid6VrklHkq
sGqDqHCdP6O8TTB5dDDItllVo6_1pcbUrhiUSMxbbXUvdvWXzg-
UD8biiReQFlfz28zGWVsdiNAUf8ZnyPEgVFn442ZdNqiVJRmBqrYRXe8P_ijQ7p8Vdz0TTrxUeT
3lm8d9shnr2lfJT8ImUjvAA2Xez2Mlp8cBE5awDzT0qI0n6uiP1aCN_2_jLAeQTlqRHtfa64QQS
UmFAAjVKPbByi7xho0uTOcbH510a6GYmJUAfmWjwZ6oD4ifKo8DYM-X72Eaw",
    "consumerSignature":
"eyJhbGciOiJQUzM4NCIsImtpZCI6ImJpbGJvLmJhZ2dpbnNAaG9iYml0b24uZXhhbXBsZSJ9.S
XTigJlzIGEgZGFuZ2Vyb3VzIGJ1c2luZXNzLCBGcm9kbywgZ29pbmcgb3V0IHlvdXIgZG9vci4g
WW91IHN0ZXAgb250byB0aGUgcm9hZCwgYW5kIGlmIHlvdSBkb24ndCBrZWVwIHlvdXIgZmVldCw
gdGhlcmXigJlzIG5vIGtub3dpbmcgd2hlcmUgeW91IG1pZ2h0IGJlIHN3ZXB0IG9mZiB0by4.cu
22eBqkYDKgIlTpzDXGvaFfz6WGoz7fUDcfT0kkOy42miAh2qyBzk1xEsnk2IpN6tPid6VrklHkq
sGqDqHCdP6O8TTB5dDDItllVo6_1pcbUrhiUSMxbbXUvdvWXzg-
UD8biiReQFlfz28zGWVsdiNAUf8ZnyPEgVFn442ZdNqiVJRmBqrYRXe8P_ijQ7p8Vdz0TTrxUeT
3lm8d9shnr2lfJT8ImUjvAA2Xez2Mlp8cBE5awDzT0qI0n6uiP1aCN_2_jLAeQTlqRHtfa64QQS
UmFAAjVKPbByi7xho0uTOcbH510a6GYmJUAfmWjwZ6oD4ifKo8DYM-X72Eaw"
  }
}
```

Returns the agreement by agreement id

`GET /get_pricing_model/{agreement_id}`

Retrieve an agreement's pricing model
pricingModel
Example data
Content-Type: application/json


```
{
  "pricingModel": {
    "paymentType": "one-time purchase",
    "price": 125.68,
    "currency": "$",
    "fee": 6.28,
    "paymentOnSubscription": {
      "timeDuration": "string",
      "repeat": "string"
    },
    "isFree": false
  }
}
```


Returns the pricing model by agreement id

`GET /check_active_agreements`

Retrieve all the active agreements. (The agreements become active when they are created and stored on the blockchain.)
Returns a list of active agreements

`GET /check_agreements_by_consumer/{consumer_public_keys}`
`/{active}`

Retrieve all or just the active agreements of a consumer
Path parameters

- consumer_public_keys (required)
- active (required)

Example data


```
-   [
      {"kty":"EC","crv":"P-
      256","x":"6MGDu3EsCdEJZVV2KFhnF21xCRI5yNpf4vWQrCIMk5M","y":"0OZbKA
      dooCqrQcPB3Bfqy0g-Y5SmnTyovFoFY35F00M","alg":"ES256"}
    ]
-   false
```


Return type
Returns all/active agreements based on consumer's public keys

`GET /check_agreements_by_provider/{provider_public_keys}`
`/{active}`

Retrieve all or just the active agreements of a provider

Path parameters

- provider_public_keys (required)
- active (required)

Example data


```
-   [
      {"kty":"EC","crv":"P-
      256","x":"4sxPPpsZomxPmPwDAsqSp94QpZ3iXP8xX4VxWCSCfms","y":"8YI_bv
      VrKPW63bGAsHgRvwXE6uj3TlnHwoQi9XaEBBE","alg":"ES256"}
    ]
-   true
```


Return type

Returns all/active agreements based on provider's public keys

`GET /check_agreements_by_data_offering/{offering_id}`

Retrieve all agreements for a data offering

Returns all agreements by offering id

`GET /retrieve_agreements/{consumer_public_key}`

Retrieve the active agreement by consumer public key whose start date is reached

Returns active agreement by consumer public key whose start date is reached

`GET /state/{agreement_id}`

Check the state of the agreement: active, violated, or terminated

Returns agreement's state based on agreement id

`POST /evaluate_signed_resolution`

Evaluate a signed resolution

body signed_resolution (required)


```
  {
  "proof":
"eyJhbGciOiJFUzI1NiJ9.eyJwcm9vZlR5cGUiOiJyZXNvbHV0aW9uIiwiZGF0YUV4Y2hhbmdlS
WQiOiJTTmg5eUtYYjJlaGxWSFJZQkllay16Z1pVaDJtU1NvMWpwbGg3SWEtNHlRIiwiaWF0Ijox
NjQ2OTUxNjM1LCJpc3MiOiJ7XCJhbGdcIjpcIkVTMjU2XCIsXCJjcnZcIjpcIlAtMjU2XCIsXCJ
kXCI6XCJ1Z1NpSTlJTEdnTWM1TmMwbkFhM3FGTjNBTjBvR2JhMzNJQWFrSHFkdm1nXCIsXCJrdH
lcIjpcIkVDXCIsXCJ4XCI6XCJMNldmVlhHYkgwaW82SnBtOTRTMWxwZGk2eUd0VDFPbVo2NUFfa
1NfaGs4XCIsXCJ5XCI6XCI2WUUwb1BPcFdCcUM3NURfanRKVWZ5NWxzWGxHak81ZzZRWGl2RHdN
REtjXCJ9Iiwic3ViIjoie1wiYWxnXCI6XCJFUzI1NlwiLFwiY3J2XCI6XCJQLTI1NlwiLFwia3R
5XCI6XCJFQ1wiLFwieFwiOlwiVlhzQnVPWndWamhvZkpWNGtBaGJhNnduMUVZRHdVSWtnWGIyZl
ZuTDh4Y1wiLFwieVwiOlwiaDRmTDVRdjRFWXQ3WGRLcWRJeTFaSnM0X1FXWURrWTF6VXpTb0k2M
U43WVwifSIsInJlc29sdXRpb24iOiJkZW5pZWQiLCJ0eXBlIjoiZGlzcHV0ZSJ9.TtxUm3E6Lfm
wEI74cr6RO4-nw-xcFaeARYOZ4z1dBVlc_JU0mCv0Ftr9tCDhggfLiJqb4RIPiNfIytFZMUbx-
g",
  "sender_address": "0x4d82Bd33baA4Fe5489C45bBdC206019403dcF829"
}
```

Returns a raw transaction for the create agreement operation

```
POST /propose_penalty
```

Propose penalty
Request body
body choose_penalty (required)


```
{
  "agreementId": 15,
  "chosenPenalty": "NewEndDateForAgreementAndReductionOfPayment",
  "paymentPercentage": 16,
  "newEndDate": 189898999
}
```


Returns the chosen penalty and sends notification to the provider with the chosen penalty

```
PUT /enforce_penalty
```

Agree to penalty by enforcing it on the blockchain
Request body
body enforce_penalty (required)


```
{
  "senderAddress": "0xC6b8cf76BD7078e56C6CE8C357dD91caeEa70170",
  "agreementId": 15,
  "chosenPenalty": "NewEndDateForAgreementAndReductionOfPayment",
  "paymentPercentage": 16,
  "newEndDate": 189898999
}
```


Returns a raw transaction for the enforce penalty operation

```
PUT /terminate
```

Terminate agreement for batch data based on the last block of successful transfer and for streaming data if the end date is reached
body terminate (required)


```
{
  "senderAddress": "0xC6b8cf76BD7078e56C6CE8C357dD91caeEa70170",
  "agreementId": 15,
  "proof": "JWT",
}
```


Returns a raw transaction for the terminate agreement operation

Explicit consent:

```
POST /give_consent
```

Give consent to a user
body consent (required)

```
{
    "dataOfferingId": "63909dae0863a775a4d71bc9",
    "consentSubjects": [

     "did:ethr:i3m:0x026b23ab3cc76f1da1d5d2aa087d29894146ee52b56c23392a7f1
     36f7dc2a7a90c",

     "did:ethr:i3m:0x020bc2643908df0e6ab258a2dac38cd3b42ce2088a0a4e3b501d4
     85ababf9f5ad6",
    ],
    "consentFormHash":
  "36bede32098bd09e15a23274a37117e58a8b08bf54a1e48331a1ff8cc509e6da",
    "startDate": 1633344669,
    "endDate": 1673344669,
    "senderAddress": "0x9aDA42ff81B9D661cC4fdab62791DaC30cfe7305"
}
```


Returns a raw transaction for the give consent operation

`PUT /revoke_consent`

Revoke consent by consent subjects
body consent (required)


```
{
    "dataOfferingId": "63909dae0863a775a4d71bc9",
    "consentSubjects": [

     "did:ethr:i3m:0x026b23ab3cc76f1da1d5d2aa087d29894146ee52b56c23392a7f1
     36f7dc2a7a90c"
    ],
    "senderAddress": "0x9aDA42ff81B9D661cC4fdab62791DaC30cfe7305"
}
```


Returns a raw transaction for the enforce penalty operation

`GET /check_consent_status/{dataOfferingId}`

Retrieve consent status
Returns a list of consent status based on data offering and consent subject (optional)

`POST /deploy_consent_signed_transaction`

Deploy signed transaction and send encrypted notification based on the event emitted by the ExplicitUserConsent smart contract
body signed_transaction (required)
Returns transaction receipt with confirmation 1

7.6 Background Technologies

- **Hyperledger BESU:**

1	Technology
Technology name	**Hyperledger BESU**
Summary	Hyperledger BESU is an Ethereum client designed to be enterprise-friendly for both public and private permissioned network use cases. It can also be run on test networks such as Rinkeby, Ropsten, and Görli. Hyperledger BESU includes several consensus algorithms including PoW and PoA (IBFT, IBFT 2.0, Etherhash, and Clique). It also supports features including privacy and permissioning.
Description	Hyperledger BESU is an open-source Ethereum client developed under the Apache 2.0 license and written in Java. It runs on the Ethereum public networks, private networks, and test networks such as Rinkeby, Ropsten, and Görli. BESU implements Proof of Work (Ethash) and Proof of Authority (IBFT 2.0 and Clique) consensus mechanisms. BESU includes a command line interface and JSON-RPC API for running, maintaining, debugging, and monitoring nodes in an Ethereum network. BESU nodes support authentication and authorization, that is, identifying the user that performed the API query and allowing the execution of a specific set of methods. BESU supports two authentication mechanisms: username and password or JWT public key; see Figure 7.10. The communications are performed using the API via RPC over HTTP or via WebSockets. The API supports typical Ethereum functionalities such as: • ether mining; • smart contract development; • decentralized application (Dapp) development. The resultant BESU architecture is the following: 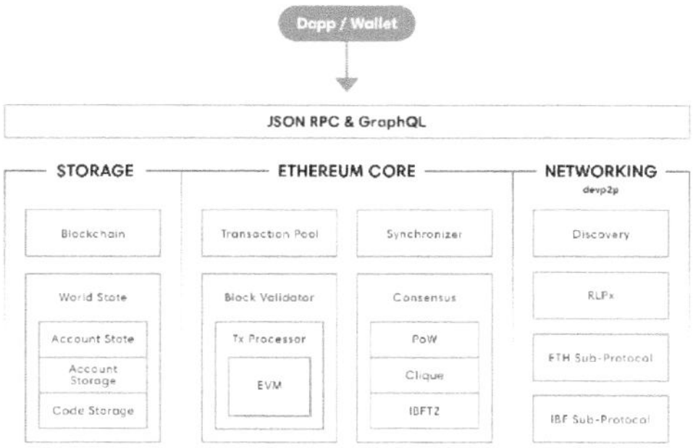 **Figure 7.10** BESU architecture.

1	**Technology**
	BESU uses a private transaction manager, Orion, to implement privacy. Each BESU node sending or receiving private transactions requires an associated Orion node. Private transactions pass from the BESU node to the associated Orion node (see Figure 7.11). The Orion node encrypts and directly distributes (that is, point-to-point) the private transaction to the Orion nodes participating in the transaction. 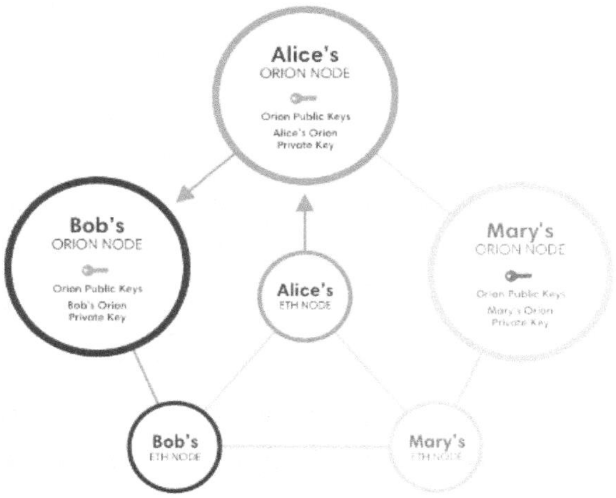 **Figure 7.11** Alice sends a private transaction to Bob using Orion privacy manager. BESU also supports permissioning, which stands for permitting only specified nodes and accounts to participate by enabling node permissioning and account permissioning on the network. It supports local permissioning (a configuration file for each node) or on-chain (via smart contracts).
Keywords	Blockchain, distributed ledger, Ethereum, privacy, permissioning, authentication

ICT problem(s) and related functionality(ies)	Bullet list of the ICT problem(s) that the technology solves and associated functionalities.
	• **Distributed ledger**
	  ○ Auditable data storage
	  ○ Persistent transaction history
	  ○ Permissioned and non-permissioned network
	  ○ Pseudo-anonymous user identity
	• **Smart contracts**
	  ○ Turing-complete machine
	  ○ Immutable code (auditable and verifiable)
	• **Privacy**
	  ○ Send cryptocurrency using private transactions
	  ○ Execute smart contracts using private transactions

<table>
<tr><th>1</th><th>Technology</th></tr>
<tr><td></td><td>
• Authentication
o JWT-based tokens
o Username and password
o JWT public key authentication
• Monitoring
o Visual representation of declining node or network performance
o Collection of log files to enable issue diagnosis
• Communications
o Full-nodes and miners using HTTP/WebSockets
o Encrypted communications for privacy (Orion) and signer (Eth-Signer) using TLS
</td></tr>
<tr><td>TRL</td><td>Current technology readiness level of the technology:
• TRL 7 – system prototype demonstration in operational environment</td></tr>
<tr><td>Website</td><td>https://www.hyperledger.org/projects/besu</td></tr>
<tr><td>Standards</td><td>BESU nodes are compatible with Ethereum public network. It supports different consensus protocols: Proof of Work (Ethash) and Proof of Authority (IBFT 2.0 and Clique).
The communications use HTTP and JSON-RPC protocols. Clients can be authenticated using JWT.
Smart contracts are coded using Solidity.</td></tr>
</table>

- **Solidity:**

Solidity is an object-oriented, high-level language for implementing smart contracts. Smart contracts are programs that govern the behaviour of accounts within the Ethereum state. It is a curly-bracket language. It is designed to target the Ethereum virtual machine (EVM).

Solidity is used to develop the smart contracts that are deployed on the Ethereum blockchain.

- **Hardhat:**

Hardhat is a development environment for Ethereum software. It consists of different components for compiling, debugging, and deploying smart contracts, all of which work together to create a complete development environment.

Hardhat has a plug-in for integration with ethers.js, which is a compact library for interacting with the Ethereum blockchain.

- **Swagger:**

Swagger is a set of open-source rules and tools for developing RESTful APIs. It simplifies the process of writing APIs by specifying the standards and providing the tools required to write safe, performant, and scalable APIs. Moreover, the Swagger framework allows developers to create interactive, machine and human-readable API documentation.

8

i3-MARKET Crypto Token and Data Monetization

8.1 Objectives

The federation of independent data spaces/marketplaces further calls for a highly secure, trusted, and cost-efficient payment solution.

At first a standard payment solution has been designed providing a protocol to exchange data with a non-repudiable and auditable accounting of data transfers. This ensures transparent billing and support for conflict resolution.

This protocol is based on a cryptographic proof exchange between data consumer and data provider and a final recording of this proof on the blockchain as "notarization" on the data exchange.

Then a tokenization solution has been designed providing a crypto token based on Ethereum standard ERC-1155 and the concept of "distributed treasury", which means that each data marketplace joining the federation could exchange token for fiat money with a fixed value. The tokens minted by each marketplace are "tagged" differently so that there is always a link between the tokens and the issuer, which must provide the associated amount of fiat money during a "clearing" phase between the data marketplaces.

This allows instant currency exchange among all the stakeholders participating in the federation and also supports full audibility of all transactions.

However, until the landscape of cryptocurrencies and tokens is clarified, with the EU Parliament vote on adopting MiCA(https://eur-lex.europa.eu/legal-content/EN/TXT/?uri=CELEX%3A52020PC0593) regulation, which is expected to establish harmonized rules for crypto-assets at the EU level, thereby providing legal certainty and guidance as to the usage of crypto tokens, the i3-MARKET Alliance decided to only use tokens as a means for distributing fees for the long-term sustainability of the system.

The choice to establish a new crypto token for real-time trading of data assets between federated data spaces and marketplaces in the i3-MARKET platform was made to overcome the boundaries of individual marketplaces and build a trustworthy working environment, one of the keys to establishing a single European data economy.

In particular, the problems solved by the adoption of the token as a currency (like the use of the blockchain in our non-repudiable protocol) are:

- to exchange value in a peer-to-peer manner, without the need for someone in between;
- to make sure that ownership is transferred and that information about the exchange cannot be tampered with.

This is extremely disruptive for the market because information is decentralized and the control is distributed among all network marketplaces, thus avoiding the designation of an impartial central intermediary. Decentralized communities provide certainty of identity, certainty of provenance and the smart contracts, like the treasury smart contract designed here, and certainty of execution ("if I pay with tokens, I get value in return") in the network.

The main technical contributions are the design and development of an *ad-hoc* blockchain-based non-repudiation protocol and the design of a crypto token solution based on the concept of "distributed treasury" to allow and trust the real-time trading of data assets among federated data spaces and marketplaces.

The designed non-repudiable protocol, instead of relying on trusted third parties for storing the non-repudiation evidence information, uses the Hyperledger BESU blockchain deployed on the i3-MARKET nodes that preserve both the proof of origin and proof of receipt of the parties involved. This is to ensure two things: one is that the information sent cannot be denied, for example, a DC has sent a message to a DP, so the DC cannot deny the behaviour. The other is that the recipient of the information cannot be denied. Similarly, DP has sent a message to DC, but DC cannot claim that it has not received this message. In this sense, the blockchain guarantees both the non-repudiation of information and the non-tampering of data by the parties involved.

The tokenization component and the cryptographic token flow have been designed to support all the different interactions between the subjects of i3-MARKET, namely data marketplace, data provider, and data consumer. To enable the trading of assets among the i3-MARKET network marketplaces, a custom flow has been designed consisting of four different phases, which, starting from the mint of the token from a data marketplace, allow it to be used as a means of payment for data or fees and to trace the path up to the return to the original marketplace, where it is burned. In particular, the use of the token in these phases enables payments that are easy to make, reliable, safe, and verifiable both within and between marketplaces and allows correct management to differentiate the tokens issued by different marketplaces.

8.2 Technical Requirements

For the components of the data monetization subsystem, the following requirements have been defined in the form of epics and user stories.

Standard payments:

Epics:

Name	Description	Labels
Standard payments	Standard payments refer to payments for a specific dataset or piece of data. Initially, it should support: payment in advance – *a posteriori* payment. In both cases, it should also support: pay per dataset or specific piece of data – subscription (flat rate within a specific set of conditions)	Epic data Consumer Data provider Data marketplace

User stories:

Name	Description	Labels
In advance payment	As a data provider, I want to be paid in advance for providing my data so that I can monetize them immediately	User story Data provider Data consumer Data marketplace
***A posteriori* payment**	As a data provider, I want to be paid *a posteriori* for providing my data so that I can have more consumers to subscribe my offering	User story Data marketplace Data consumer Data provider
Non-repudiation Protocol	As a data provider, I want to provide my data with a Non-repudiation Protocol so that I can bill data consumers based on reliable data exchanges As a data consumer, I want to consume data with a non-repudiation protocol so that I can contest wrong billings	User story Data consumer Data provider

Tokenization:

Epics:

Name	Description	Labels
Currency tokenization	The federation of independent data spaces/marketplaces further calls for a highly secure, trusted, and cost-efficient payment solution. Therefore, a suitable crypto currency solution that allows instant currency exchange among the participating data spaces/marketplaces and also supports full audibility of all transactions has to be provided	Epic Data marketplace Data consumer Data provider Data owner

User stories:

Name	Description	Labels
Provide crypto tokens (exchange in)	As a data marketplace, I want to provide crypto tokens to data consumer so that I will enable P2P payments for data exchange As a data consumer, I want to purchase crypto tokens from a data marketplace so that I can subscribe offering from other marketplaces in the i3-MARKET network	User story Data marketplace Data consumer
Payment with crypto tokens	As a data provider, I want to receive payment with crypto tokens so that I can receive instant payments	User story Data marketplace Data consumer Data provider Data owner
Withdraw crypto tokens (exchange out)	As a data provider/data owner, I want to receive fiat currency from a data marketplace so that I can monetize the crypto tokens received for providing my data As a data consumer, I want to receive fiat currency from a data marketplace so that I can monetize my crypto tokens if I leave the i3-MARKET network	User story Data marketplace Data consumer Data provider Data owner
Clearing	As a data marketplace, I want to receive fiat currency for the tokens emitted by other data marketplaces so that I can monetize these tokens if I leave the i3-MARKET network or with a specific scheduling	User story Data marketplace

8.3 Solution Design/Blocks

The following are the high-level capabilities provided by the data monetization subsystem:

1. Standard payments: In advance or *a posteriori* payment for a specific dataset or piece of data.
2. Tokenization: Creation of a crypto token solution for instant currency exchange among the participating data spaces/marketplaces.
3. Pricing manager: Managing of i3-MARKET cost and price model.

From Figure 8.1, the data monetization subsystem block interacts with following two building blocks:

- Data storage system: The data monetization subsystem uses the data Storage system for recording crypto token transactions.
- Backplane system: The data monetization subsystem has been used from the Backplane system for accounting and executing payment operations for data purchases and tokenization operations.
- Data access system: The data monetization subsystem has been used from the data access system for accounting and/or executing payments for data exchanges.

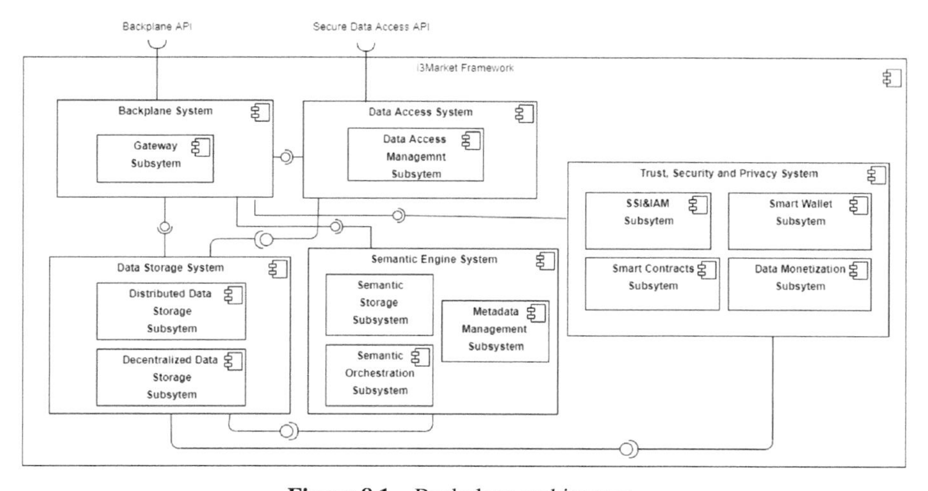

Figure 8.1 Backplane architecture.

See Figure 8.2 for the *specific component diagram.*

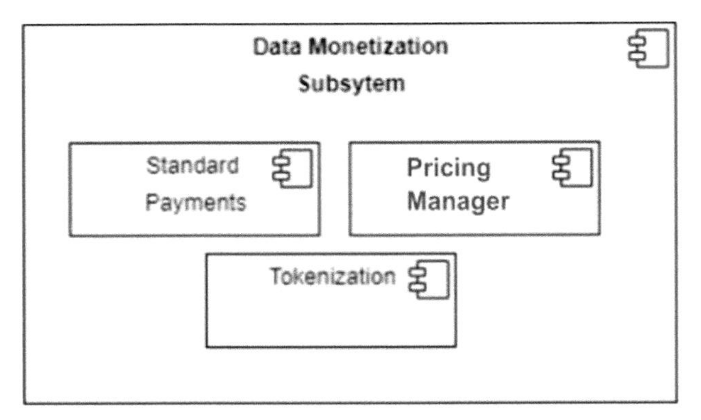

Figure 8.2 Data monetization components.

The data monetization subsystem is in charge of providing "standard payments", "pricing manager", and "tokenization" capabilities.

Inside, we can find:

- component "standard payments" responsible for managing the payments, in advance or *a posteriori*, for a specific dataset or piece of data;
- component "tokenization" responsible for the creation of a crypto token for instant currency exchange and other tokenization operations among the participating data spaces/marketplaces;
- component "pricing manager" responsible for managing the price and the cost model.

8.4 Standard Payment

The Non-repudiation Protocol aims at preventing parties in a data exchange from falsely denying having taken part in that exchange. The protocol flow begins when a data consumer requests a block of data from the data provider. See Figure 8.3 for details.

At first, the data provider has to generate a one-time symmetric secret key (JWK) for a given JWA algorithm identifier; this secret is going to be used to encrypt the data block required by the data consumer (Figure 8.4). After the data encryption, the data provider builds a proof of origin JWT object, containing information about the parties (source, destination, etc.), the timestamp, the hash algorithm used, the hash of the block, the secret key, and the encrypted cipherblock. See Figure 8.5 for more details.

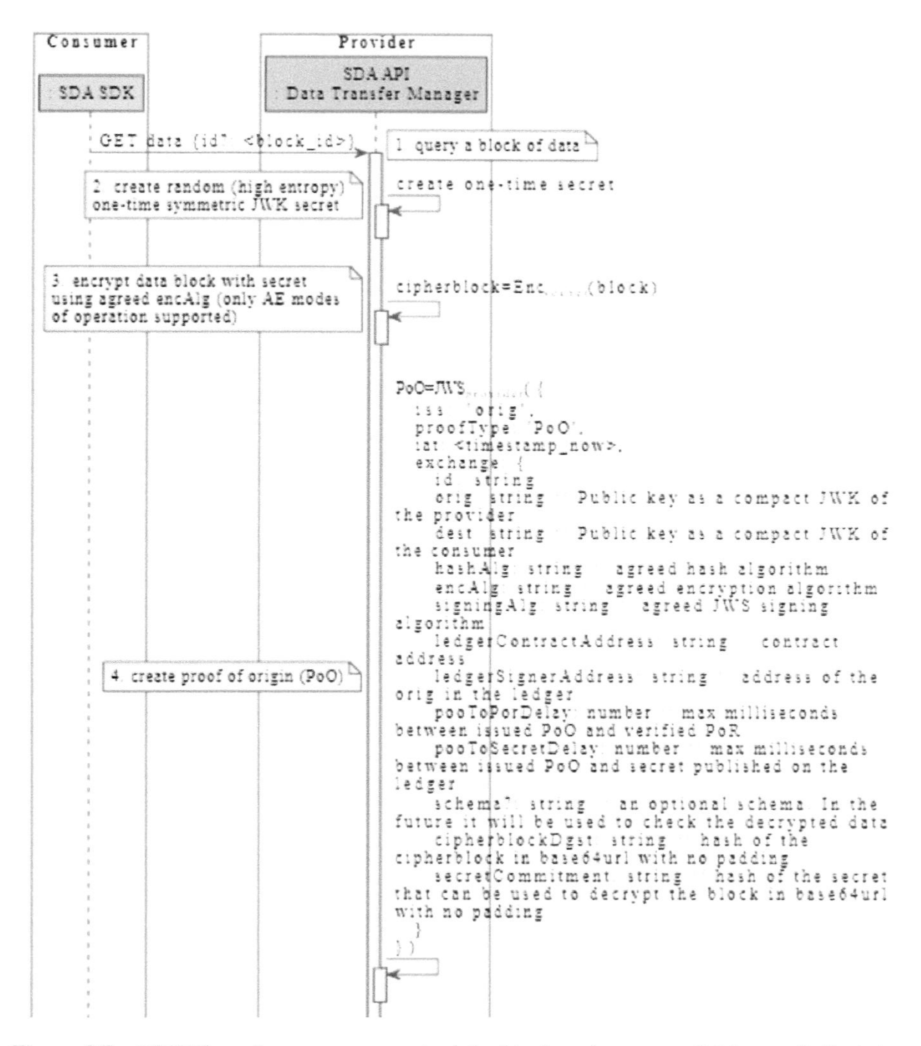

Figure 8.3 NRP Phase 1 — consumer gets cipherblock and non-repudiable proof of origin.

This object is then signed with the private key of the data provider and returned to the data consumer.

The data consumer, at this point (Figure 8.5), can validate the proof received using the data provider public key. If the validation is successful, he can store the proof in his local memory. After having completed these steps, he generates another proof, the proof or receipt. This proof is generated as

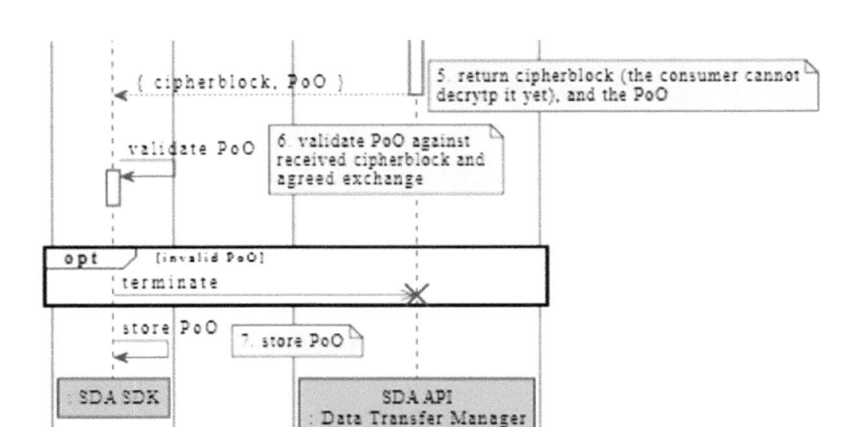

Figure 8.4 NRP Phase 1 Part 2.

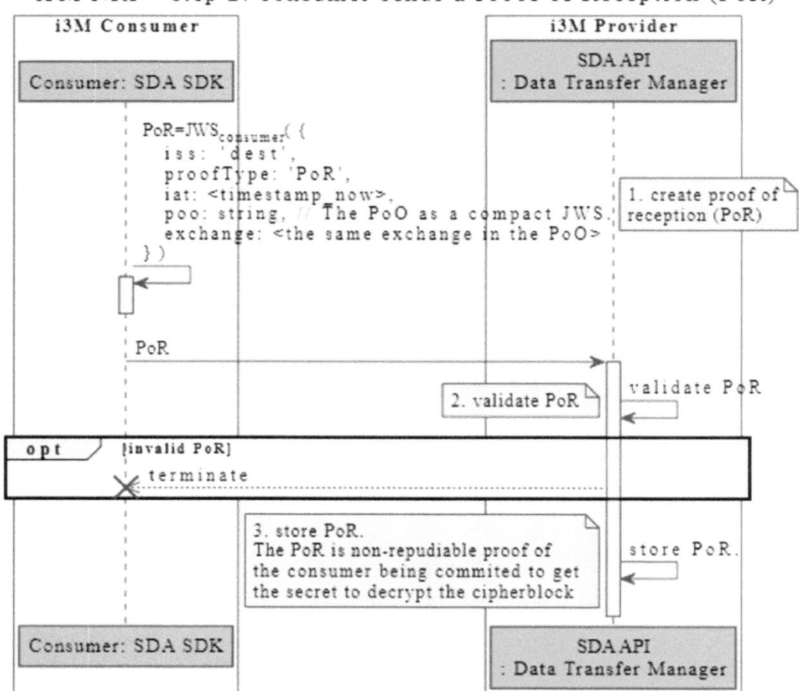

Figure 8.5 NRP consumer sends PoR.

another JWT object containing information about the parties (iss, sub, etc.), the timestamp, the hash of the received proof of origin, and the hash algorithm used.

The proof of receipt is then signed with the data consumer private key and sent back to the data provider. Once proof of receipt is received, the data provider can validate it using the data consumer public key; see Figure 8.6.

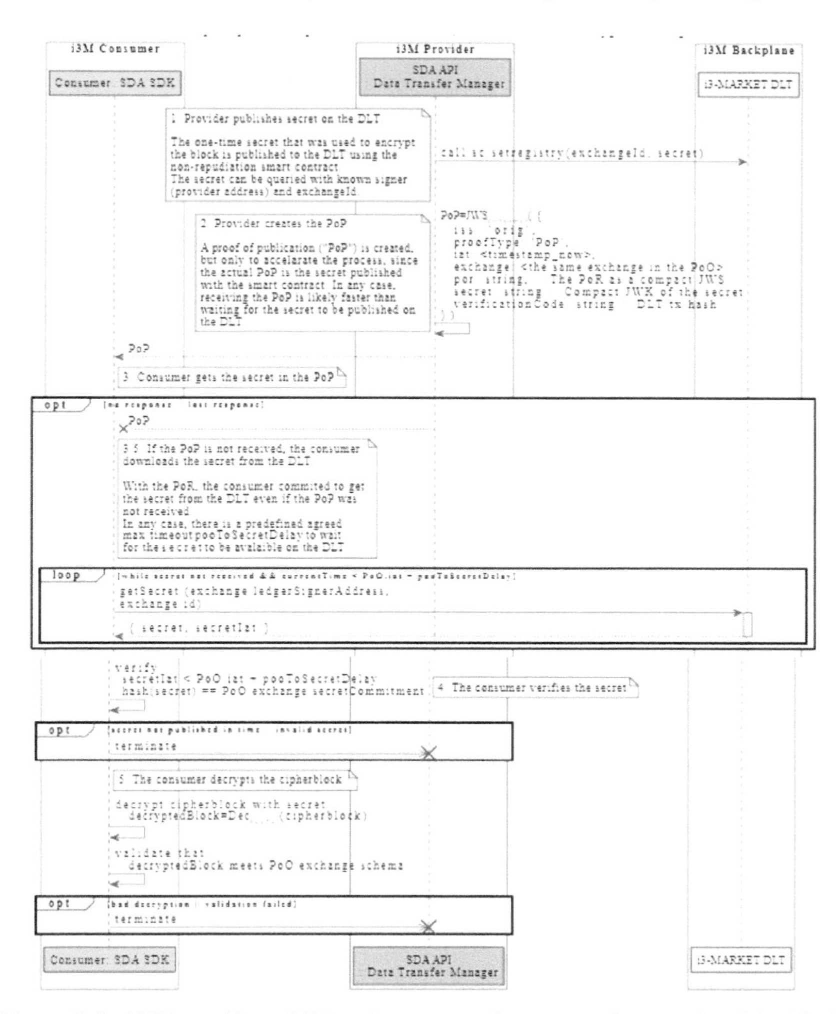

Figure 8.6 NRP provider publishes the secret, and consumer decrypts the cipherblock.

The provider now publishes the one-time secret that was used to encrypt the block on the i3M BESU blockchain. A proof of publication is then created but only to accelerate the process, since the actual proof of publication is the secret published within the smart contract. The PoP is then sent to the data consumer.

If the data consumer does not receive the proof and the key in a predefined/agreed max timeout, he can retrieve them directly from the blockchain. Once having received the proof of publication and the secret key, the data consumer can validate the proof of publication with the auditable accounting public key and verify that the hash of the key received is equal to the hash of the key previously received in the proof of origin.

As the last step, if the verification is successful, he can decrypt the block with the secret key and validate it with the hashed block included in the proof of origin. If some validation or verification problems arise, the flow will enter a conflict resolution phase.

8.5 Tokenization

The tokenization process and the components used to create and manage a cryptographic token for instant currency exchange and other tokenization operations among participating i3-MARKET actors are represented in Figure 8.7.

i3-MARKET tokenization architecture:

Starting from the right of the architecture in Figure 8.21, to manage the operational flows between the various data spaces/marketplaces involved in

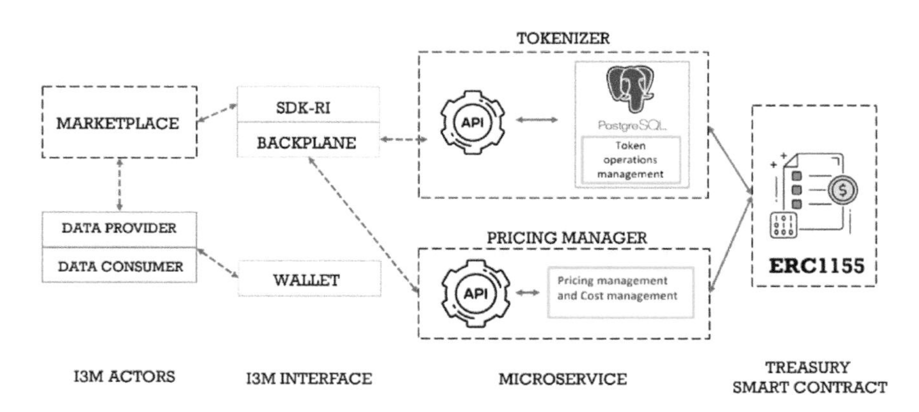

Figure 8.7 Tokenization process.

currency exchange within the i3-MARKET platform, a specific i3-MARKET treasury smart contract has been created. This smart contract contains and maintains for each wallet the token balance of the data marketplaces, DP, DC, and community members. More specifically, it is responsible for managing the secure transfer of tokens between the parties and for tracking immutably payments made in both tokens and fiat money.

To enable interaction with the treasury smart contract functionalities, we have created two microservices, the tokenizer and the pricing manager. The tokenizer allows the i3-MARKET actors to interact with the treasury smart contract and keep track of all the marketplace operations, and the pricing manager manages the data price and the fees. These two services are integrated with the i3M Backplane and the SDK-RI to be used from an i3-MARKET DM.

Treasury smart contract operations:

The most important features involving the tokenization operations are presented below; these have been implemented within the treasury contract, which extends the ERC-1155 standard.

- **Register a data marketplace:**

To register a new data marketplace and its token type, a mapping to bond the data marketplace addresses and the index identifier of the new token type is required.

The function that inserts a new marketplace in the smart contract, increments an index variable and is added in the mapping of the marketplace address as key and using a unit value as the identifier of the new token type.

```
contract I3-MARKETTreasury is ERC1155 {

    uint public index = 0;
    mapping(address => uint) public mpIndex;

    constructor() public ERC1155("https://i3-MARKET.com/marketplace/{id}.json"){
    }

    /*
    * add a new Data Marketplace in the platform
    */
    function addMarketplace(address _mpAdd) external onlySameAdd(_mpAdd)
onlyNewMpAdd(_mpAdd) {
        index += 1;
        marketplaces.push(_mpAdd);
        mpIndex[_mpAdd] = index;
    }
}
```

- **Exchange in:**

The exchange method must be called by a data marketplace, which issues and transfers the right amount of tokens (of its token type) to the user who pays in fiat money.

ERC1155 function:

_mint(address account, uint256 id, uint256 amount, bytes data)

- **Fee payment:**

The payment method should transfer the token fees, taken "arbitrarily" from the token types available in the data consumer balance, to the data provider.

The ERC1155 function used for the payment:

```
safeBatchTransferFrom(address from, address to,
uint256[] ids, uint256[] amounts, bytes data)
```

Example: Starting from the first token type in the balance loop until the amount is covered.

```
function configurePayment(address from, uint256 amount) private view returns
(uint256[] memory ids, uint256[] memory amounts) {
uint256[] memory mpIds = new uint256[](index);
        uint256[] memory mpTokens = new uint256[](index);
        for (uint256 i = 0; i < index && amount != 0; ++i) {
            uint256 mpBalance = super.balanceOf(from, i + 1);
            if (mpBalance != 0) {
                mpIds[i] = i + 1;
                mpTokens[i] = getMarketplaceNeededTokens(mpBalance, amount);
                amount = amount - mpTokens[i];
            }
        }
        require(amount == 0, "NOT ENOUGH TOKENS");
        return (mpIds, mpTokens);
}
```

- **Exchange out:**

The *exchange-out* method should transfer the right amount of token, taken "arbitrarily" (first the tokens belonging to the data marketplace in the exchange out and once finished, the others) from the token types available in the balance, from a community member to a data marketplace.

The ERC1155 function used for the *exchange-out* operation:

```
safeBatchTransferFrom(address from, address to,
uint256[] ids, uint256[] amounts, bytes data)
```

- **Clearing:**

The clearing method should be called for every token type present in the data marketplace balance, aside from the token type the data marketplace has created.

The ERC1155 function used for the clearing operation:

```
safeTransferFrom(address from, address to,
uint256 id, uint256 amount, bytes data)
```

```
struct ClearingOperation{
    string transferId;
    address toAdd;
    uint tokenAmount;
}

function clearing(ClearingOperation[] memory _clearingOps) external payable
onlyMp(msg.sender){

    //clearing for each marketplace contained
    for (uint i = 0; i < _clearingOps.length; ++i){
        isMarketplace(_clearingOps[i].toAdd,"ADD ISN'T A MP");
        if(_clearingOps[i].tokenAmount > minimumClearingThreshold) {
            super.safeTransferFrom(msg.sender,_clearingOps[i].toAdd,mpIndex[_clear
ingOps[i].toAdd], _clearingOps[i].tokenAmount, "0x0");

            //create transaction with isPaid param to False as Fiat money payment
is not completed yet
            txs[_clearingOps[i].transferId] =
TokenTransfer(_clearingOps[i].transferId, msg.sender, _clearingOps[i].toAdd,
_clearingOps[i].tokenAmount, false, "");
            emit TokenTransferred(_clearingOps[i].transferId, "clearing",
msg.sender, _clearingOps[i].toAdd);
        }
    }
}
```

- **Exchange-out and clearing strategy:**

Since a marketplace has first to collect the fiat money from all the DM involved in an exchange-out operation before transferring the money requested, here we describe the suggested strategy that each marketplace should implement in its code. Other strategies can be used in agreement with the network marketplaces.

Requirements:

- Set a number variable **X** as the interval of days that a marketplace collects exchange-out requests.

- Set a numeric variable **Z** as the interval of days in which a marketplace must wait for other marketplaces to pay in fiat money for the tokens sent via clearing.

Strategy flow:

Marketplace ordered steps for the exchange-out operation:

1. Starting from the first day a marketplace starts operating, in the first **X** days, the marketplace should collect all the exchange-out requests from the users (community).
2. At the end of **X** days, the marketplace asks to exchange the tokens in its balance that belong to other marketplaces.
3. Now the marketplace should wait for another **Z** days so that all the other marketplaces can pay with fiat money the tokens sent with the clearing operation.
4. Once the **Z** days have passed and the fiat money from the clearings are collected, the marketplace can pay out the users that in the first **X** days requested the exchange-out of tokens.

The marketplace can restart in parallel this flow and collect another round of exchange-out requests at the end of point 2.

Tracking of token transfers (exchange out, clearing):

The token payment process involves storing in the blockchain the history of the transactions made once the token transfer is completed.

We want to save the transfer operation in a mapping of structs, where the key identifier is a unique value generated outside.

```
// object that stores the token transfer information
struct TokenTransfer {
    uint transferId;
    address fromAddress;
    address toAddress;
    uint tokenAmount;
    bool isPaid; //True if the fiat money payment has been completed
                 //False if only the token transfer is completed
    string transferCode;

}

//mapping to track all the token transfer transactions
mapping(uint => TokenTransfer) public transactions;
```

Tokenizer service:

The main purpose of the tokenizer service is to allow the i3-MARKET actors (i.e., marketplace, data provider, and data consumer) to call the i3-MARKET treasury smart contract methods and interact with the i3-MARKET token flow; see Figure 8.8. The tokenizer is a Node.js backend service with a local Postgres database to persist the marketplace token activities.

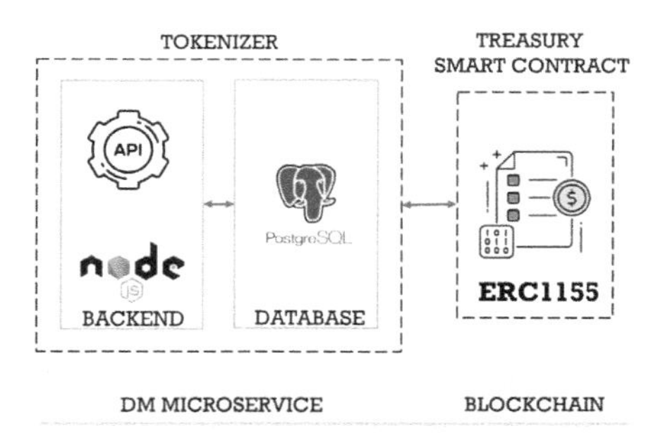

Figure 8.8 Tokenizer architecture.

Each data marketplace needs an instance of the tokenizer and a dedicated local database. The tokenizer tracks the status of transactions made by a marketplace on the treasury smart contract using blockchain events, while the full history of each transaction is stored in the local database for verification and error prevention.

To deploy a transaction on the i3M BESU blockchain, the transaction should be first signed with the i3-MARKET Wallet. The tokenizer *post* operations create a transaction object that must be signed and then deployed separately using the flow presented below.

1. The first step is to create a new raw transaction using one of the *post* operations available (i.e., exchangeIn, exchangeOut, clearing, fee-payment, etc.). After a successful transaction, the payload of the response will be a transaction object like this one:

```
{
    "transferId":"68aa8652-6457-5786-81cb-2ee2cc906aa6",
    "transactionObject":{
        "nonce":12,
        "gasLimit":12500000,
        "gasPrice":204695,
        "to":"0x3663f8622526ec82aE571e4265DAd6967dd74260",
        "from":"0x50c0F1E9ACF797A3c12a749634224368ebC1f59A",
        "data":"0x909770870000000000000000000000000000000000000000870000a18943
    }
}
```

With this operation, the marketplace tokenizer service saves in its database the operation with the operation_name (i.e., exchangeIn, exchangeOut, clearingIn, clearingOut, fee-payment, etc.) with status *UNSIGNED-OPERATION*, the address of the user involved, the date, and a unique transferID to get this operation at a later time:

TRANSFERID	OPERATION	STATUS	USER	DATE
1111	operation_name	unsigned_operation	address	date

2. Now the raw transaction has to be signed with the i3M wallet. Only the fields contained in the "transactionObject" are used for the signing transaction operation. Below is the object to be signed:


```
{
    "nonce":12,
    "gasLimit":12500000,
    "gasPrice":204695,
    "to":"0x3663f8622526ec82aE571e4265DAd6967dd74260",
    "from":"0x50c0F1E9ACF797A3c12a749634224368ebC1f59A",
    "data":"0x90977087000000000000000000000000000000000870000a1894332
}
```


3. The next step is to deploy the signed transaction. Once the marketplace gets the signed raw transaction, it can call the deployment endpoint of the tokenization service /treasury/transactions/deploy-signed-transaction. The response of the request should be a long transaction object with information about the transaction.
4. When the operation deployment is successful, the marketplace tokenizer service updates in its database the previous operation with status open.

TRANSFERID	OPERATION	STATUS	USER	DATE
1111	operation_name	open	address	Later-date

Pricing manager service:

Pricing manager is a Java microservice to configure and evaluate the price and the cost of data; see Figure 8.9. The microservice uses the i3-MARKET BESU blockchain and an in-memory database to persist data.

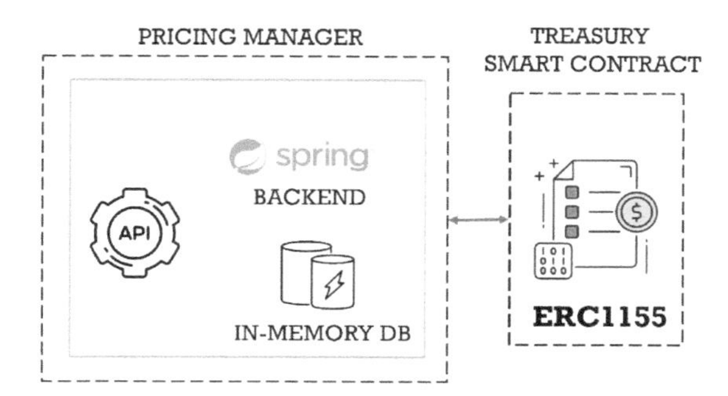

Figure 8.9 Pricing manager architecture.

The service APIs are logically divided into two subsets.

Pricing management:
This service allows to calculate the price of some data based on a preconfigured formula. The service, through the exposed APIs, allows you to manage the formula and customize the parameters and constants.

The formula and the constant values are stored in an in-memory database inside the service, as every marketplace can have a customized formula if needed.

Currently, in the i3-MARKET platform, all the data marketplaces will use the formula provided by AUEB.

Cost management:
This service can be used to calculate the fee of some data, which depends on the price of the data and the percentage of the fee. The fee percentage is stored in the blockchain and more specifically in the treasury smart contract.

8.6 Diagrams

The following diagrams describe the processes involving the components of the data monetization subsystem.

These requirements have been collected using the Trello Boards system taking into account functional use cases and general requirement reported by partners, stakeholders, big companies, and SMEs.

Standard payment:

A Non-repudiable Protocol is used for accounting data transfers. Based on the accounted data exchanges and smart contract information about data consumer (company name, VAT, billing address, etc.) and pricing, the data provider will invoice the data consumer; see Figure 8.5. The payment will be done using standard bank payment methods.

- **Accounting:**

The Non-repudiation Protocol aims at preventing parties in a data exchange from falsely denying having taken part in that exchange. To ensure the traceability of data exchanges and manage conflicts, the proof of origin of the data provider and the proof of receipt of the data consumer are stored in the immutable ledger; see Figure 8.10.

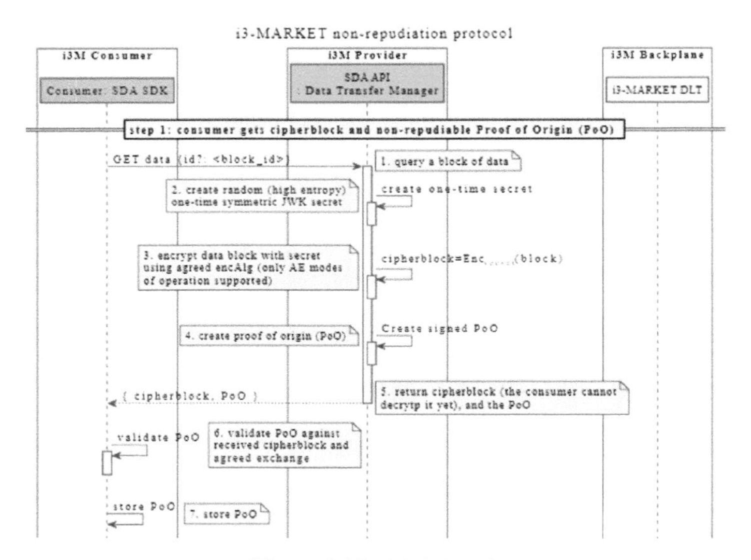

Figure 8.10 NRP Part 1.

In the first step of the protocol, shown in Figure 8.11, a block of data is requested from the data consumer to the data provider. The data provider encrypts the requested block, creates the proof of origin, and returns this proof to the consumer with the encrypted block. At this point, the data consumer validates the proof of origin received from the data provider, stores the proof in its local memory, and creates the proof of receipt.

In the second step, the data consumer sends the proof to the data provider; see Figure 8.11. The data provider validates the proof of receipt and stores it in local storage. As a third step, the provider publishes the secret on the blockchain and sends the proof of publication to the data consumer.

The data consumer can obtain the proof of publication and the cryptographic key directly from the data provider or from the blockchain if this is not received within a maximum time; see Figures 8.12 and 8.13. The data consumer then checks that the key is received, that the PoP is valid, and that

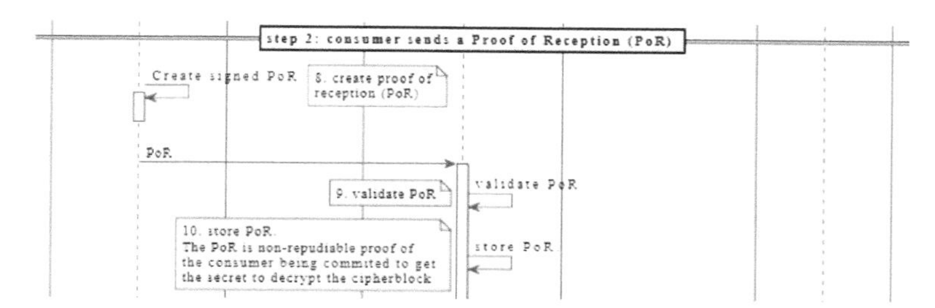

Figure 8.11 NRP Part 2.

the key hash is the same as the key commitment parameter included within the PoO. At this point, the data consumer decrypts and validates the previously received cipherblock.

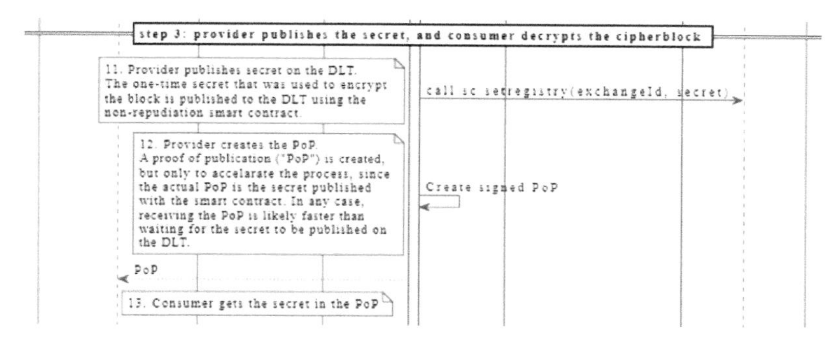

Figure 8.12 NRP Step 3 Part 1.

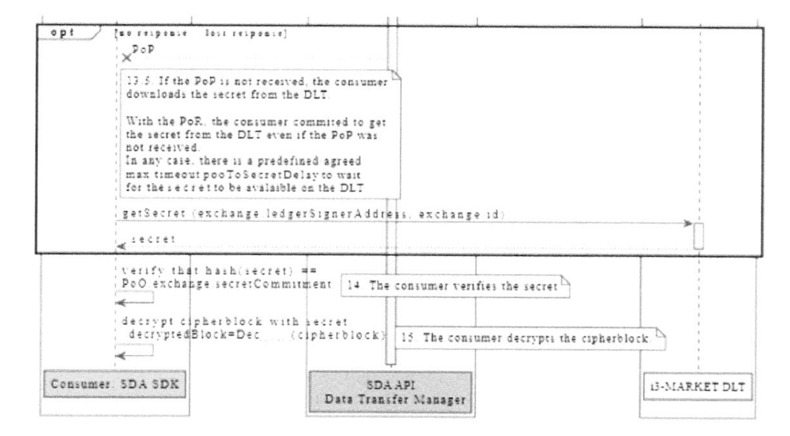

Figure 8.13 NRP Step 3 Part 2.

- **Invoice management:**

  In this process, the data provider retrieves from the Backplane the information (based on the accounted data exchanges and the smart contract agreement) to produce the invoices for the data consumers.

  Data consumers could check the invoices verifying the accounted information and pay the invoices with standard payment methods; see Figure 8.14.

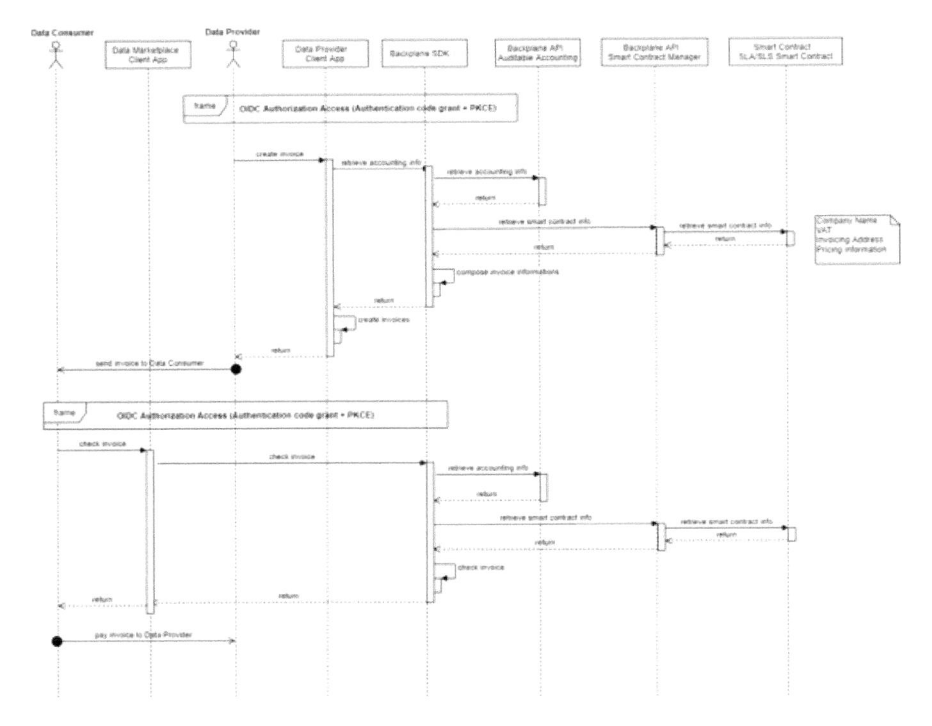

Figure 8.14 Invoicing process.

Tokenization:

An i3-MARKET crypto token has been created customizing Ethereum ERC-1155 standard. The treasury smart contract contains and maintains the different balances for each data marketplace and user in the i3-MARKET network. When a data consumer obtains tokens from a data marketplace paying fiat money (*exchange in*), both the total balance of data consumer wallet and the specific data marketplace balance will be increased.

During the *payment for data* phase between a data consumer and a data provider, the data consumer can pay the data price in fiat money or in tokens to the data provider. In addition to the data price payment, the data consumer will pay some fees in tokens to the i3M community, the provider data marketplace, and the consumer data marketplace; see Figure 8.15.

A community member (or a DP if we enable data price payments in tokens) will be able to ask fiat money for his token balance from any of the network DM (exchange out) and the amount of token will be transferred from the total balance of community member wallet to the balance of DM wallet; the community member can pay with tokens belonging to the DM with which it is doing the operation or with tokens belonging to other DMs.

Finally, a DM will be able to ask for the clearing of tokens distributed by the other DMs (clearing) – Figure 8.15. For each specific DM balance of the requesting DM wallet, a clearing request will be created, and all the DMs involved will be notified and should pay fiat money to the requesting DM and confirm clearing execution. On clearing confirmation, the tokens will be transferred from requesting DM wallet to clearing DM wallet (requesting DM already approved the transfer during the clearing request) in Figure 8.16.

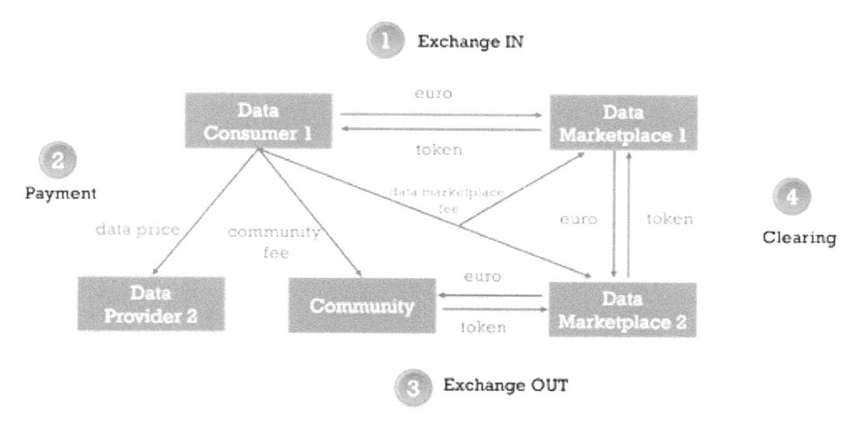

Figure 8.15 Tokenization model.

- **Exchange in:**

In the *exchange in* phase, the user requests a specific amount of tokens from a data marketplace, which, upon receiving the payment in fiat money from the user, returns the tokens in the amount requested as depicted in Figure 8.16.

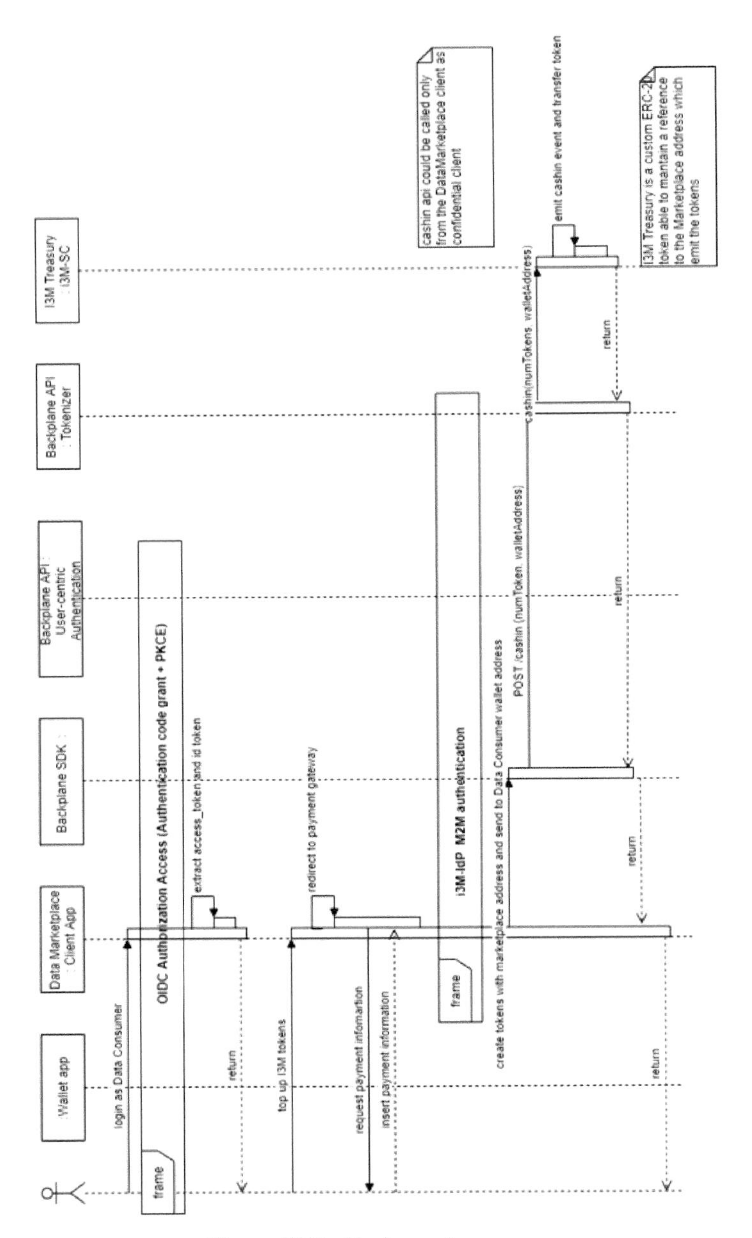

Figure 8.16 Exchange in process.

As shown in the flow above, when the fiat money payment is received by the data marketplace, it authenticates with the Backplane so that it can call the "exchange in" method of the treasury smart contract; see Figure 8.17. The smart contract mints the tokens directly into the address of the user who requested the tokens; see Figure 8.17.

- **Payment:**

In the *payment* phase, first, the user logs in as data consumer and then can start a data exchange. As a first step, the data consumer makes a first

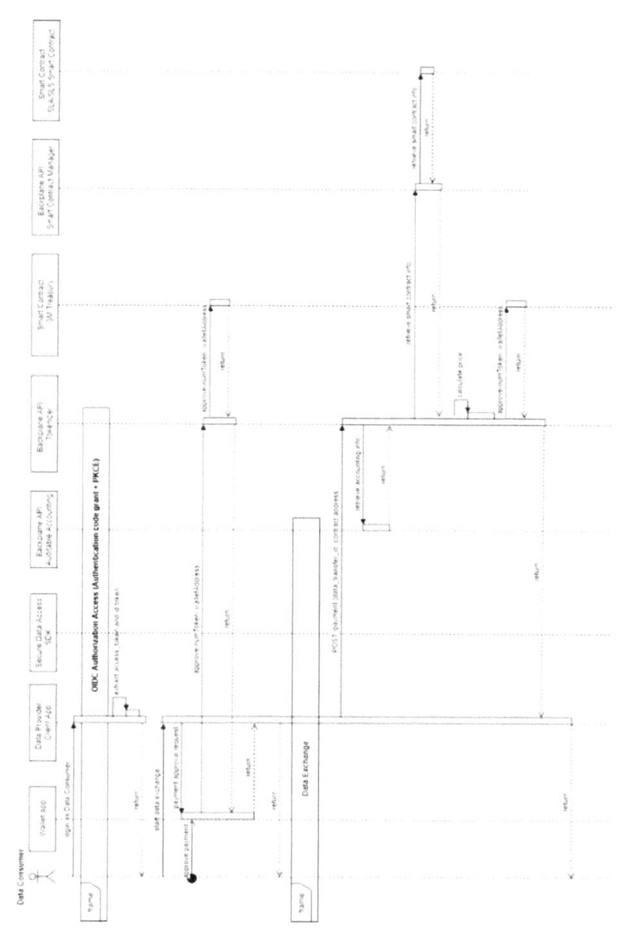

Figure 8.17 Payment process.

transaction in the blockchain in which he inserts tokens on deposit and, therefore, commits himself to the data provider by offering a guarantee (monetary security). When the data exchange is concluded, the right amount of tokens are taken from the data consumer deposit and moved to the data provider balance; see Figure 8.18.

- **Exchange out:**

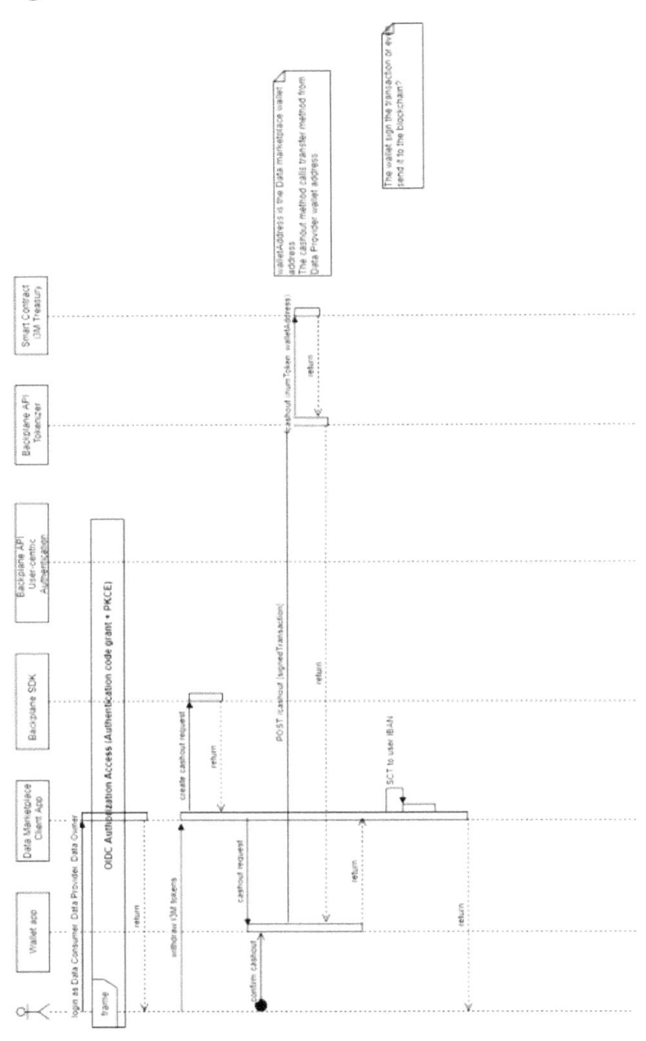

Figure 8.18 Exchange out process.

In the *exchange out* phase, the data provider after the login with the data marketplace can start the withdrawal of the i3-MARKET tokens in his balance sheet (Figure 8.19). At first, he confirms the cash-out operation via his wallet application that calls the "exchange out" method in the treasury smart contract. If the operation is successful, the smart contract saves the information regarding the token transaction. At this point, the data marketplace proceeds with the payment in fiat money to the user IBAN and publishes the transaction identifier (TRN) in the blockchain to securely store the transaction in case of future conflicts.

- **Clearing request:**

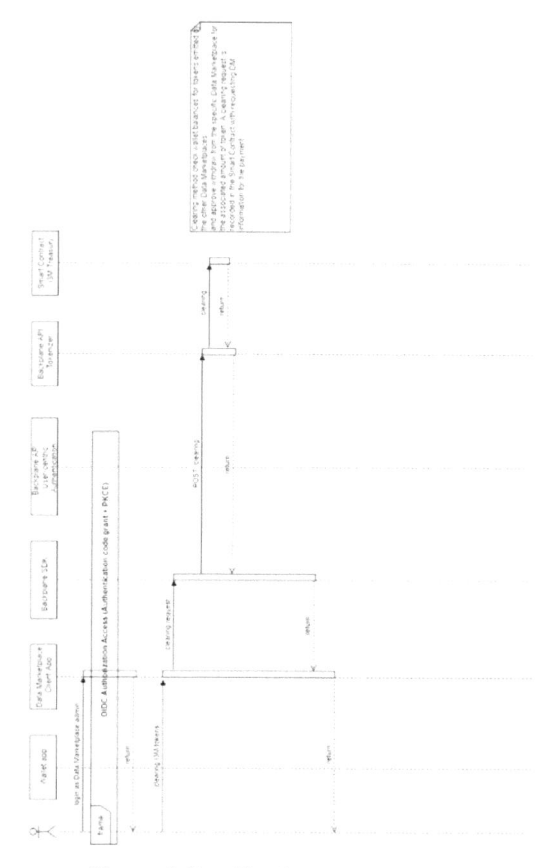

Figure 8.19 Clearing request process.

During the *clearing* phase, the data marketplace that wants to leave the i3-MARKET network should return the tokens in its balance to the corresponding data marketplace owners – see Figure 8.20. The clearing method should be called for every token type present in the data marketplace balance aside from the token type the data marketplace has created.

- **Clearing execution:**

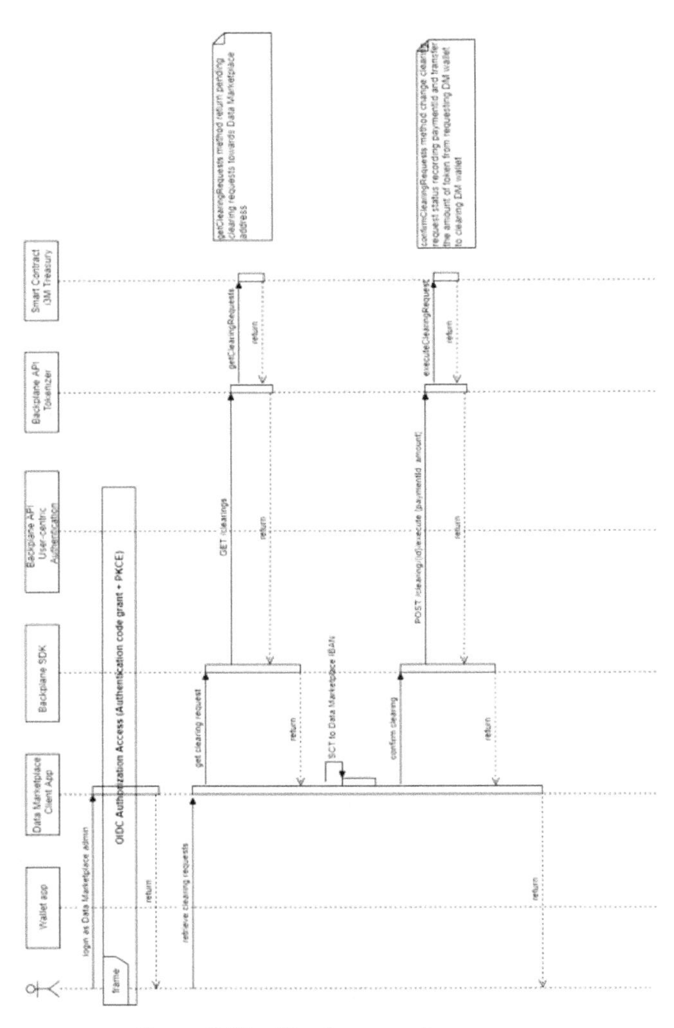

Figure 8.20 Clearing execution process.

In the *clearing execution* phase, the data marketplaces that receive a clearing request from a data marketplace must pay the corresponding value in fiat money of the tokens received.

8.7 Interfaces

The interfaces of the library of the Non-repudiation Protocol for standard payment of the treasury smart contract for tokenization and pricing manager microservices are presented below – see Figures 8.21 and 8.22.

Tokenization:

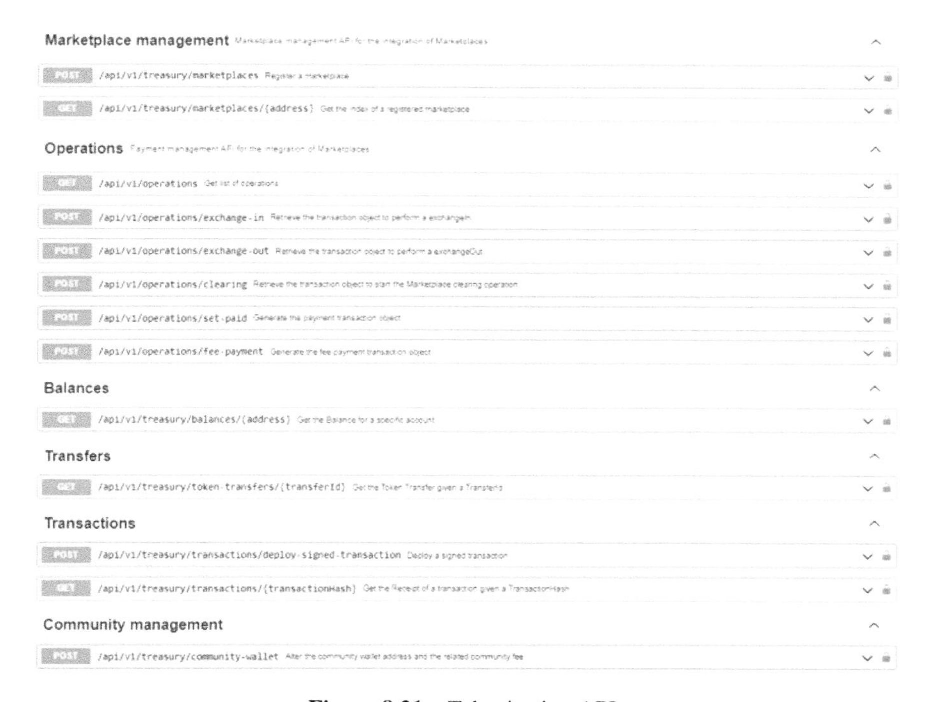

Figure 8.21 Tokenization API.

Pricing manager:

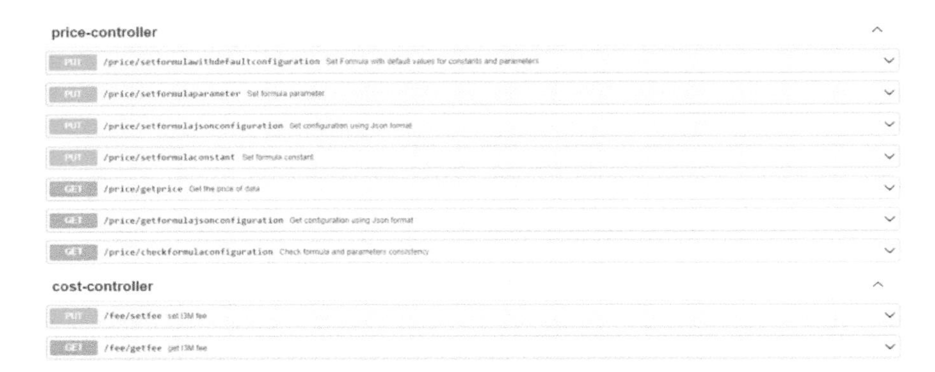

Figure 8.22 Pricing manager API.

8.8 Background Technologies

To implement the solution for tokenization, the ERC-1155 multi-token standard has been chosen and customized.

The ERC-1155 standard was used to implement the i3-MARKET treasury contract, which outlines a smart contract interface that can represent any number of fungible and non-fungible token types. More specifically, the ERC-1155 multi-token standard allows each token ID to represent a new configurable token type, which can have its own metadata, supply, and other attributes.

Adapted to our solution, this token standard has been used to collect the token for the different marketplaces that adhere to the platform, where a new fungible token has to be created for every new marketplace that joins the consortium. This solution allows us to track at any time which token type and therefore which marketplace the tokens in the balance sheet for any participant in the network belong to.

This is particularly important because, for example, during a clearing operation, a marketplace should know to which different marketplaces its tokens in the balance sheet belong; therefore, it can send to each one the right amount of tokens to be converted and returned in fiat money.

It is also important to underline the advantages that the 1155 standard brings because in token standards like ERC-20 and ERC-721, a new separate contract has to be deployed for each token type or collection. This places a lot of redundant bytecodes on the Ethereum blockchain and limits certain functionalities by the nature of separating each token contract into its own permissioned address. With this new design, it is possible to transfer multiple token types at once, saving on transaction costs and removing the need to "approve" individual token contracts separately.

9

i3-MARKET Semantic Model Repository and Community

The results are shared not only with project partners but also with stakeholders and community in open-source repositories. As part of open-source assets, the data models, documentations, and files used in the i3-MARKET project are made available, such as the following:

- The i3-MARKET *data pack* is the set of files, schemas, and metadata model diagrams that represent the way the i3-MARKET semantics is organized and structured; it also contains the metadata in two different formats, e.g., ttl and Jason-ld. owl.
- The i3-MARKET *semantic model* info is the documentation that describes in detail all the taxonomies and vocabularies from needed domains used in i3-MARKET and that describes and represents all the relationships between them to build the graph representation of the i3-MARKET semantic model.
- The *support* repo is the mechanism for how the data model is maintained following the interoperability requirements in i3-MARKET. If you want to contribute or have any suggestion for improving the semantic models, visit the open-source repositories and contact authors and members.
- The *model files are shared in i3-MARKET GitHub/Gitlab repositories* with release versions where each section contains the online machine-readable files in OWL and other formats for online accessibility. The files are maintained and updated regularly to keep the latest version of the model files up to date.

The code as well the models and vocabularies are available open-source via the establishment of the i3-MARKET spaces on Gitlab available at: https://gitlab.com/i3-MARKET-V3-public-repository/ and GitHub available at: https://github.com/i3-MARKET-V3-public-repository/.

i3-MARKET semantic model governance process, which is defined as the support and evaluation process to include semantic improvements, is as follows:

- **Request for changes or updates:** Identify any changes prior to a *major* release, which should be considered private and usually is on testing and pre-consensus/staging.
- **The evaluation of any type of update request:** A review from editors and community, approves participation, and updates. In particular terms, vocabularies, ontologies or initiate a model extension in the i3-MARKET OSS project.
- **The communication of the results from technical experts:** A tagging version using alpha, beta, and gamma versions and then tagged as major is used here.
- **Evaluation of contributions for new commits:** Technical experts, PM, TM, TPMs, WPLs, and TaskLs assess and evaluate the contribution, including documentation at the initiated project in i3-MARKET OSS.
- **Reports and changes report:** The technical board issues a short report, explaining the rational on the rejection in exceptional cases; this step can include rejecting/cancelling project participation.

It is possible to find a more complete definition of the attributes used in the data offering description schema template as used in the semantic engine API in Appendix A.

9.1 Semantic Engine (SEED)

The semantic engine and framework solution is available and integrated into the i3-MARKET Backplane. Another concept is the metadata semantic registry stored in a registry database (like MongoDB). With this feature, the Backplane can rely on the metadata registry storage capacity to collect the semantic information about the assets and information for the marketplaces and stakeholders that can be created, searched, retrieved, and manipulated for external and internal operations.

Semantic engine framework:

From an operational perspective, i3-MARKET envisages semantic engine components (e.g., SEED) to manage query mechanisms on top of the registry catalogues, including complex discovery and retrieve checks that make sure, e.g., that the necessary information is retrieved by the actors and services. Also very important are the functionalities related to the creation and registration of the data offering descriptions and the management of local and federated registries. The data offerings can be shared by providers/marketplaces in the i3-MARKET network and the engine can search, discover, and retrieve the data offerings, which are authorized, from all the nodes/marketplaces.

The engine also has functionalities and interfaces that are used in conjunction with other Backplane components/systems to compile and fill information and details for the functionalities, for example, for notification manager, smart contract manager, data access & transfer, and BESU.

Semantic engine and metadata framework:

a. Data offering creation
b. Data offering discovery
c. Data offering registry
d. Federated discovery on different instances
e. Management of data sharing agreement and service agreement parameters to comply with contract manager operations
f. Alignment with entities and IDs in Backplane information models

We developed and implemented dedicated software components for semantic engine system as SEED, which is in charge of managing the semantic metadata, descriptions, queries, discoveries, retrieving, creating, and mapping descriptions and manipulating registries, federated queries, and component interactions and interfaces. To make easier the interface and use of functionalities, we present the external operations via APIs that are more agnostic and easier to use also for non-semantic experts.

9.2 Technical Requirements

For the semantic storage, the following high-level capabilities have been defined:

1. Semantic metadata management:
   The semantic engine (SEED) relies on a local MongoDb and Hyperledger-BESU. All the information, for instance, data provider, data offerings, consumer, and querying offering, are stored as semantic data.

Name	Description	Labels
Metadata storage (MongoDb)	The registry storage (MongoDb) is responsible to store semantic data and process the queries. The storage should provide either the REST endpoint or client connector so that other components can access to the data	Epic
Spatial and text data storage	To support spatial and full-text search queries, the semantic data manager should be able to index spatial and full-text data	Epic

Name	Description	Labels
Save semantic data	As a subject, I want to save my semantic metadata so that I can query and update it later Subject: Data Consumer, Data Provider	Epic

2. Offering registration:
   The semantic engine exposes APIs to register, query, and update offering. A data provider can regist er offering, for instance, datasets and the price for data, etc.

Name	Description	Labels
Offering registration	Offering registration is a component that allows the user to manage their data offering. More specifically, it provides the following functionalities: Register the data offering – Retrieve all the offerings – Update/delete offerings – Subscribe to an offering	Epic

3. Offering discovery:
   The semantic engine exposes APIs to query the existing offerings in i3-MARKET Backplane. A data consumer can query datasets, prices for any dataset, offering, etc.

Name	Description	Labels
Offering discovery	Offering discovery is a component that allows the data consumers to search the offering data available on the marketplace. The data consumer has to specify the characteristics of the data they are looking for. The offering discovery module will then process the data request and returns a list of available offering data that meet their requirements. More specifically, the offering discovery should provide the following functionalities: Register the consumer data request – Retrieve all the data request of a consumer – Process a data request – Update/delete a data request – Subscribe to a data request	Epic

4. Vocabulary management:

Name	Description	Labels
Vocabulary management: Semantic model management	Vocabulary management is a component that is used to manage the i3-MARKET semantic data model. More specifically, the vocabulary management should provide the following functionalities: View and search the concepts of i3-MARKET – Allow the user to propose a new concept – Allow the administrator to add a new concept – Allow the administrator to update/delete an existing concept	Epic

Backlog release – features:

Name	Description	Labels
Semantic data manager: registry storage	The registry/semantic storage is responsible for storing semantic data and processing the queries. The storage should provide either the REST endpoint or client connector so that other components can access the data	Semantic Data Storage Offerings

Features:

Name	Description	Labels
Save semantic data	As a subject, I want to save my semantic metadata so that I can query and update it later Subject: Data Consumer, Data Provider	User Story

Tasks:

Name	Description	Labels
Define a semantic description template	As a data provider, I want to create a semantic description of my offering data so that I can register it to i3-MARKET. It would be desirable that the semantic engine should provide a semantic description template so that the data provider can easily fill in to register the offering data	User Story

9.3 Solution Design/Blocks

Figure 9.1 shows the final version, which is defined as:

- Components and functionalities of semantic engine and framework
- New versions of i3-MARKET semantic models

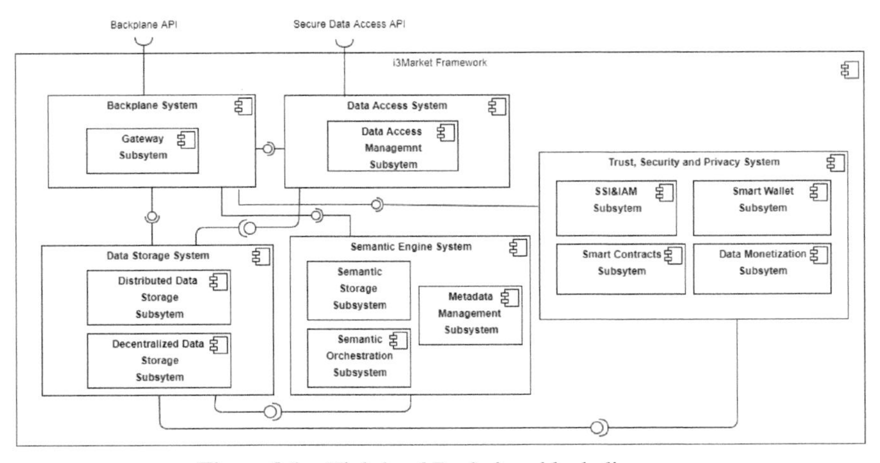

Figure 9.1 High-level Backplane block diagram.

– Semantic vocabulary management environments

Figure 9.1 shows that we use BESU SEED-INDEX library in order to retrieve all registered nodes addressed in the network and hence enabling the federated query search.

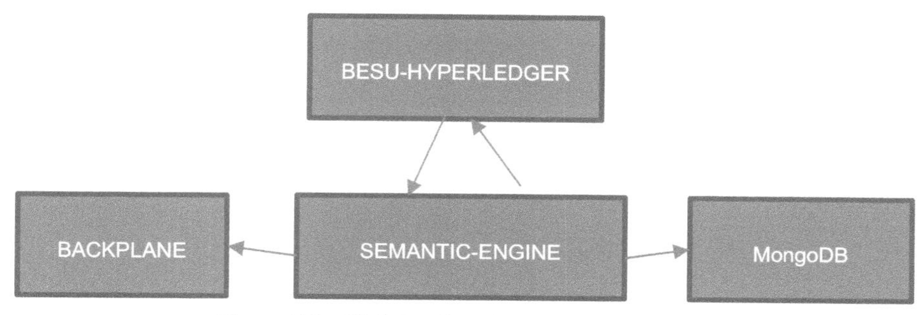

Figure 9.2 High-level Backplane block diagram.

9.4 Building Block High-level Picture

The specific component diagrams are shown in Figures 9.3–9.6.

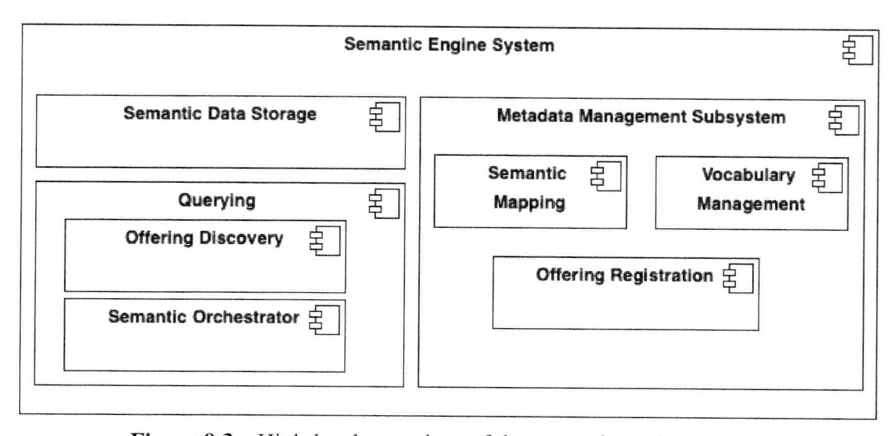

Figure 9.3 High-level operations of the semantic engine system.

For the semantic subsystems in charge of dealing with "semantic data management", we can highlight the following parts:

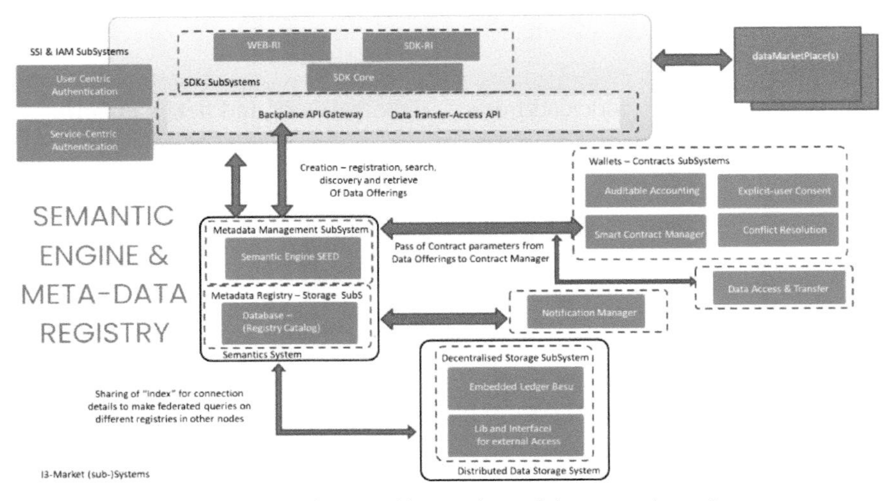

Figure 9.4 Main interfaces and interactions of the semantic engine system.

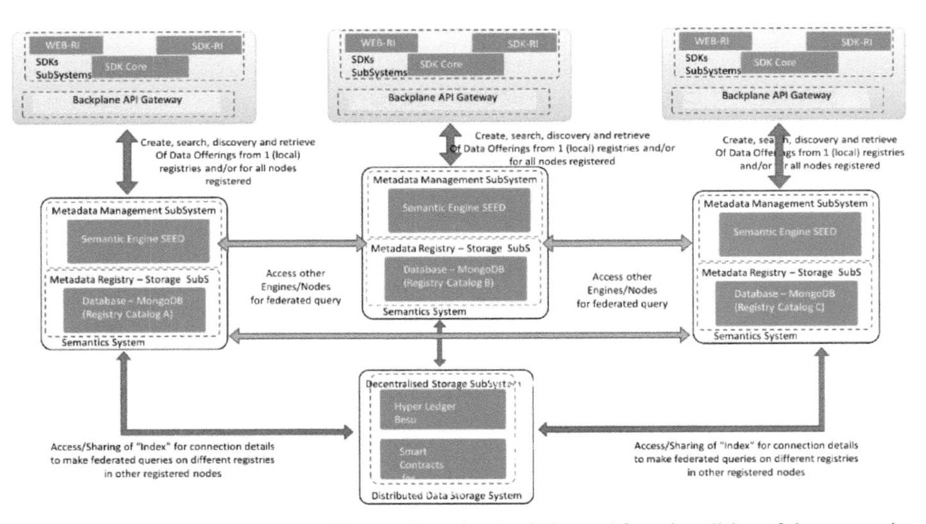

Figure 9.5 Main operations and interactions for the federated functionalities of the semantic engine system.

– Semantic data storage: This component on receiving JSON pushes the data to MongoDb database. MongoDb is a NoSql document-based database.

- Semantic mapping: This component does semantic mappings and transforms data received from API endpoints.
- Vocabulary management: This environment keeps and manages all of the vocabularies, defined as i3-MARKET semantic model, used in different operations of the semantic engine. The i3-MARKET Semantic Model is available using the GitHub and Gitlab repositories where the models/files are stored, shared, managed, and described, and the documentations is available in the developer portal.
- Offering registration: This component is basically REST APIs exposed as endpoints. Semantic engine exposes different endpoints for offering registration. Examples are:
  - register data provider;
  - register data offering of a data provider;
  - update data offerings;
  - deleting a data offering;
  - query existing offerings, etc.
- Offering discovery: This component is basically REST APIs exposed as endpoints. Semantic engine exposes different endpoints for offering discoveries and retrieving. Examples are:
  - retrieve a list of data offerings;
  - discover data offerings by providers;
  - discover data offerings by parameters;
  - discover data offerings by category;
  - discover data offerings by active state;
  - discover data offerings by shared state;
  - discover data offerings by text;
  - discover data offerings by keywords/text;
  - discover data offerings in federated search by category;
  - discover data offerings in federated search by active state;
  - discover data offerings in federated search by shared state;
  - discover data offerings in federated search by text;
  - discover data offerings in federated search by category;
  - search for particular metadata, etc.

Figure 9.6 shows a detailed landscape of the current set of microservices (cubes), API's (little yellow rectangles), components (blue rectangles), and storages (white rectangles) on i3-MARKET.

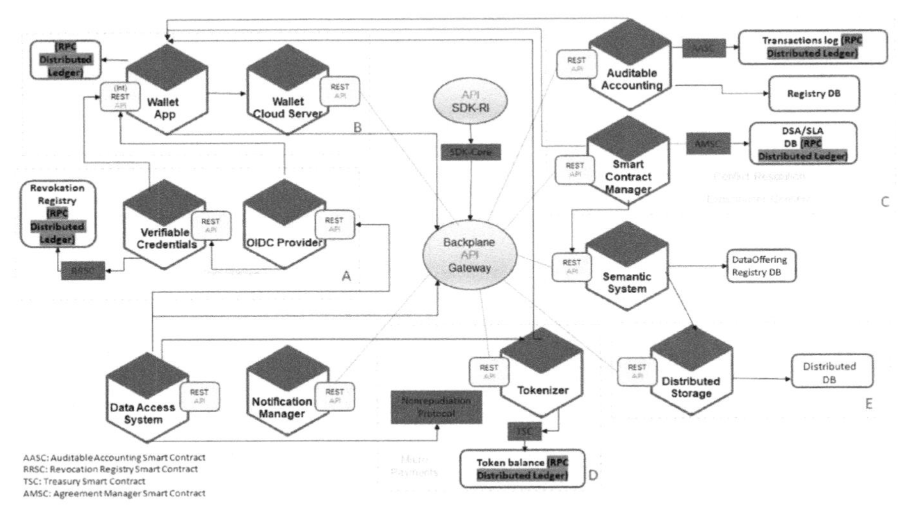

Figure 9.6 i3-MARKET services layout.

9.5 Diagrams

Data offering registration:

The diagram in Figure 9.7 shows that a data provider first must have to authenticate with i3-MARKET Backplane through a gateway. Once a provider is successfully authenticated, the provider can see all the APIs exposed by the semantic engine – called (SEED) – in the Backplane swagger interface. A provider can register an offering using registration endpoint using the template for data offering description. The Backplane internally communicates with SEED and dispatches create request to it. The engine, on receiving requests from Backplane, maps the incoming data into RDF according to the semantic data model and stores data into local registry catalogue database and sends back the response to the Backplane that offering is registered. The Backplane notifies the client/provider that offering has been successfully registered as represented in Figure 9.7.

Data offering discovery/deletion/update sequence diagram:

When a data provider interacts with i3-MARKET Backplane and has successful been authenticated, s/he can perform the following tasks:

- retrieve offering by providing offering ID;

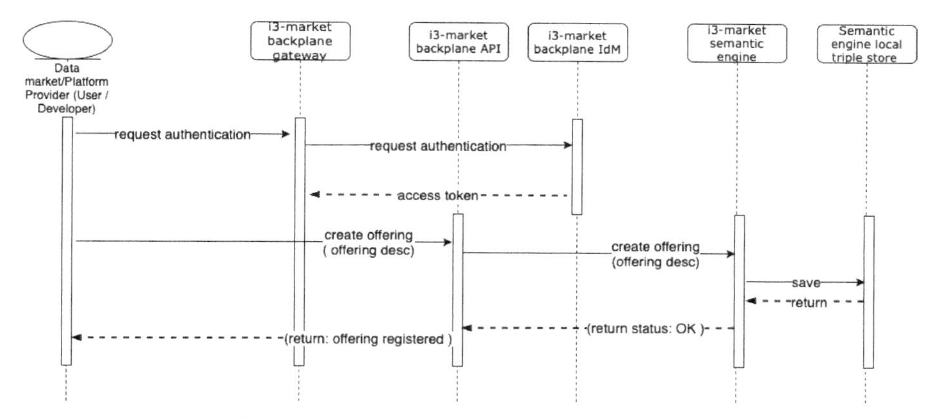

Figure 9.7 Sequence diagram for registering a data provider.

- retrieve a list of all offerings registered by a data provider using its provider ID;
- retrieve a list of all offerings filtered by category, which are registered not only in local instance of SEED (semantic engine) but also other instances of SEEDs running in the i3-Market cluster;
- update an offering;
- delete a particular data offering by providing its ID;
- the user can also download the data offering template.

The figure shows the sequence of messages used to perform different tasks. For instance, when a user wants to retrieve a particular offering s/he provides the offering ID and the Backplane sends this offering ID towards the SEED. On receiving an offering ID, the SEED executes a query on MongoDB. If the offering with the given ID is registered in the local storage, the SEED constructs the results (i.e., the requested offering) and sends back towards Backplane, where results are presented to the user. Similarly, if the user is interested to find all the offering registered by a data provider, s/he provides the provider ID and SEED looks all the offerings registered in local repository and sends back the results towards Backplane where results (list of all offerings registered by a data provider) are presented to users.

The SEED, by interacting with BESU, also can distribute the queries towards all other instances of SEEDs running on the i3-MARKET cluster. For example, if a user is interested to find offerings not only locally but also those that are registered on other instances in the i3-MARKET cluster. The

SEED engine transparently finds the offerings, filtered by category, from all the i3-MARKET instances.

Figure 9.8 shows that user can also update an offering by giving the description of an offering s/he queried by either offering by ID or data Provider ID. User has to copy the retrieved offering to the endpoint, where s/he can update any field. On receiving the updated offering, the SEED updates all the data against that particular data offering in the local storage. A data provider can also delete an offering by giving offering ID in the endpoint. Upon receiving the delete request from Backplane, the SEED executes a delete query on the local storage and the particular offering is permanently deleted from it.

To retrieve a query template, the user has to use the endpoint shown in Figure 9.8. On receiving template request, the SEED generates the offering template fully compliant with the semantic data model.

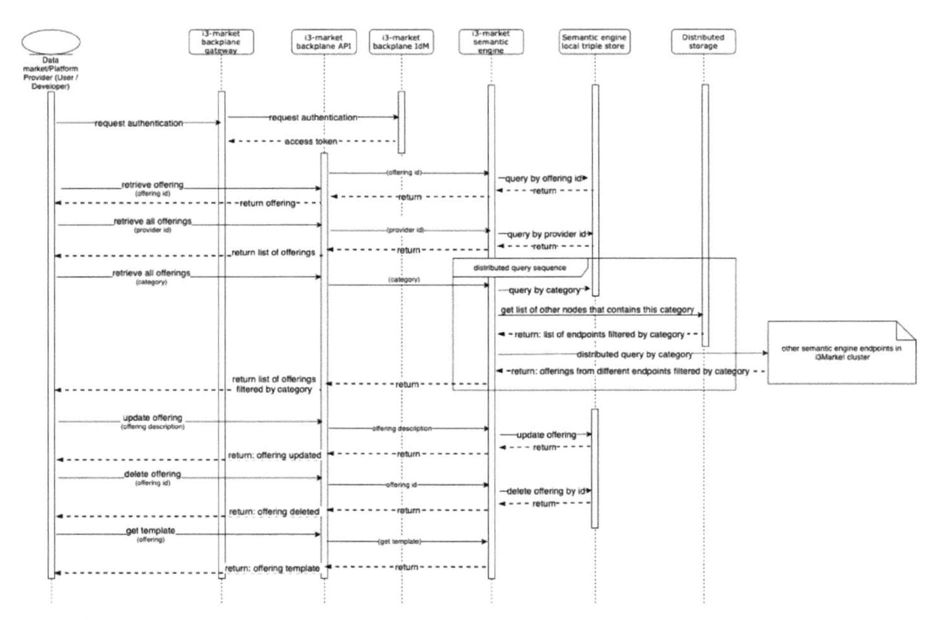

Figure 9.8 Sequence diagram for querying, deleting, and updating data offerings.

9.6 Interfaces

The semantic engine currently has many functionalities via APIs, which include: registration, searching, retrieving, updating, and deletion of different data offerings and delivery of info to other components.

Register data provider:

A data provider, when for the first-time interacts with the system, can register its information in the i3-MARKET. Following is the endpoint address and the request for registration of data provider in the semantic repository. The user must provide a "providerID" field, the ID which was provided to user at the authentication process.

```
{
  "providerId": "string",
  "name": "string",
  "description": "string",
  "organization": [
    {
      "organizationId": "string",
      "name": "string",
      "description": "string",
      "address": "string",
      "contactPoint": "string"
    }
  ]
}
```

Listing 9.1 Data provider template.

Register data offering:

When a user is registered as data provider, the next step would be to create and register data offerings (semantic descriptions) for the data assets that they want to share/sell. Below is the API pointer and the request template the user can use to register data offerings. It is important to note that the value of "provider" in the offering template should be the "ID" of the data provider.

POST /api/registration/data-offering describe data offering

```
{
  "context": {
    "core": "http://i3-MARKET.eu/Backplane/core/",
    "dcat": "http://www.w3.org/ns/dcat#",
    "pricingModel": "http://i3-MARKET.eu/Backplane/pricingmodel"
  },
  "dataOfferingId": "string",
```

```
  "provider": "string",
  "marketId": "string",
  "owner": "string",
  "providerDid": "string",
  "marketDid": "string",
  "ownerDid": "string",
  "active": true,
  "ownerConsentForm": "string",
  "inSharedNetwork": true,
  "personalData": true,
  "dataOfferingTitle": "string",
  "dataOfferingDescription": "string",
  "category": "string",
  "status": "string",
  "dataOfferingExpirationTime": "string",
  "version": 0,
  "createdAt": "2022-12-19T15:20:56.816Z",
  "updatedAt": "2022-12-19T15:20:56.816Z",
  "contractParameters": {
    "interestOfProvider": "string",
    "interestDescription": "string",
    "hasGoverningJurisdiction": "string",
    "purpose": "string",
    "purposeDescription": "string",
    "hasIntendedUse": {
      "processData": true,
      "shareDataWithThirdParty": true,
      "editData": true
    },
    "hasLicenseGrant": {
      "transferable": true,
      "exclusiveness": true,
      "paidUp": true,
      "revocable": true,
      "processing": true,
      "modifying": true,
      "analyzing": true,
      "storingData": true,
      "storingCopy": true,
      "reproducing": true,
      "distributing": true,
      "loaning": true,
      "selling": true,
      "renting": true,
      "furtherLicensing": true,
      "leasing": true
    }
  },
  "hasPricingModel": {
    "pricingModelName": "string",
    "basicPrice": 0,
    "currency": "string",
    "hasPaymentOnSubscription": {
      "paymentOnSubscriptionName": "string",
      "paymentType": "string",
      "timeDuration": "string",
      "description": "string",
      "repeat": "string",
      "hasSubscriptionPrice": 0
    },
    "hasPaymentOnApi": {
      "paymentOnApiName": "string",
      "description": "string",
      "numberOfObject": 0,
      "hasApiPrice": 0
    },
```

```
    "hasPaymentOnUnit": {
      "paymentOnUnitName": "string",
      "description": "string",
      "dataUnit": 0,
      "hasUnitPrice": 0
    },
    "hasPaymentOnSize": {
      "paymentOnSizeName": "string",
      "description": "string",
      "dataSize": "string",
      "hasSizePrice": 0
    },
    "hasFreePrice": {
      "hasPriceFree": true
    }
  },
  "hasDataset": {
    "title": "string",
    "keyword": "string",
    "dataset": "string",
    "description": "string",
    "issued": "string",
    "modified": "string",
    "temporal": "string",
    "language": "string",
    "spatial": "string",
    "accrualPeriodicity": "string",
    "temporalResolution": "string",
    "theme": [
      "string"
    ],
    "distribution": [
      {
        "title": "string",
        "description": "string",
        "license": "string",
        "accessRights": "string",
        "downloadType": "string",
        "conformsTo": "string",
        "mediaType": "string",
        "packageFormat": "string",
        "dataStream": true,
        "accessService": {
          "conformsTo": "string",
          "endpointDescription": "string",
          "endpointURL": "string",
          "servesDataset": "string",
          "serviceSpecs": "string"
        },
        "dataExchangeSpec": {
          "encAlg": "string",
          "signingAlg": "string",
          "hashAlg": "string",
          "ledgerContractAddress": "string",
          "ledgerSignerAddress": "string",
          "pooToPorDelay": 0,
          "pooToPopDelay": 0,
          "pooToSecretDelay": 0
        }
      }
    ],
    "datasetInformation": [
      {
        "measurementType": "string",
        "measurementChannelType": "string",
        "sensorId": "string",
```

```
        "deviceId": "string",
        "cppType": "string",
        "sensorType": "string"
      }
    ]
  }
}
```

Listing 9.2 Data offering template.

(When the data offering template is created, the system can use the above JSON request and store it, but the system can be updated in case to manage the data offerings as Json-ld in the registry storage.)

```
{
"@context": {
    "core": "http://i3-MARKET.eu/Backplane/core/"
    "dcat": "https://www.w3.org/ns/dcat.jsonld"
    "pricingmodel": "http://i3-MARKET.eu/Backplane/pricingmodel"
  },
  "id": "#####-#######-#######-###" OR "http://i3-MARKET.org/resource/#####-#######-#######-
###"
  "type": "http://i3-MARKET.eu/Backplane/core/DataOffering"

  "provider": "#####-#######-#######-###"
  "marketId": "#####-#######-#######-###",
  "owner": "#####-#######-#######-###",
  "dataOfferingTitle": "_field",
  "dataOfferingDescription": "string",
  "category": "Other",
  "status": "e.g. Activated, InActivated, ToBeDeleted, Deleted",
  "dataOfferingExpirationTime": "NA",
  "contractParameters":
    {
      "id": "http://i3-MARKET.org/resource/#####-#######-#######-###"
      "type": "http://i3-MARKET.eu/Backplane/core/ContractParameters"

      "contractParametersId": "string",
      "interestOfProvider": "NA",
      "interestDescription": "NA",
      "hasGoverningJurisdiction": "NA",
      "purpose": "NA",
      "purposeDescription": "NA",
      "hasIntendedUse":
        {
          "id": "http://i3-MARKET.org/resource/#####-#######-#######-###"
          "type": "http://i3-MARKET.eu/Backplane/core/IntendedUse"
```

```
        "intendedUseId": "string",
        "processData": "true OR false",
        "shareDataWithThirdParty": "true OR false",
        "editData": "true OR false"
      }      ,
    "hasLicenseGrant":
      {
        “id”: ”http://i3-MARKET.org/resource/#####-#######-#######-###”
        “type”: “http://i3-MARKET.eu/Backplane/core/LicenseGrant”

        "licenseGrantId": "string",
        "copyData": "true OR false",
        "transferable": "true OR false",
        "exclusiveness": "true OR false",
        "revocable": "true OR false"
      }
  } ,


"hasDataset":
  {
    “id”: ”http://i3-MARKET.org/resource/#####-#######-#######-###”
    “type”: “http://www.w3.org/ns/dcat#Dataset”

    "datasetId": "string",
    "title": "_field",
    "keyword": "_field",
    "dataset": "_field",
    "description": "_field",
    "issued": "date-time",
    "modified": "date-time",
    "temporal": "_field",
    "language": "_field",
    "spatial": "_field",
    "accrualPeriodicity": "_field",
    "temporalResolution": "_field",
    "distribution": [
      {
        “id”: ”http://i3-MARKET.org/resource/#####-#######-#######-###”
        “type”: “http://www.w3.org/ns/dcat#Distribution”

        "distributionId": "string",
        "title": "_field",
```

```
        "description": "_field",
        "license": "_field",
        "accessRights": "_field",
        "downloadType": "_field",
        "conformsTo": "_field",
        "mediaType": "_field",
        "packageFormat": "_field",
        "accessService":
          {
            "id": "http://i3-MARKET.org/resource/#####-#######-#######-###"
            "type": "http://www.w3.org/ns/dcat#DataService"

            "dataserviceId": "string",
            "conformsTo": "_field",
            "endpointDescription": "_field",
            "endpointURL": "_field",
            "servesDataset": "_field",
            "serviceSpecs": "_field"
          }
      }
    ],
    "datasetInformation": [
      {
        "id": "http://i3-MARKET.org/resource/#####-#######-#######-###"
        "type": "http://i3-MARKET.eu/Backplane/core/DatasetInformation"

        "datasetInformationId": "string",
        "measurementType": "_field",
        "measurementChannelType": "_field",
        "sensorId": "_field",
        "deviceId": "_field",
        "cppType": "_field",
        "sensorType": "_field"
      }
    ],
    "theme": [
      "_field"
      "_field"
      "_field"
    ]
  }
}
```

Query a registered data offering by offering ID:

Figure 9.9 shows the endpoint to fetch a particular offering registered in store. A data provider must provide the "offering ID". This offering can further be used in other endpoint (i.e., /semantic-engine/api/registration/update-offering) if the user wants to update this offering.

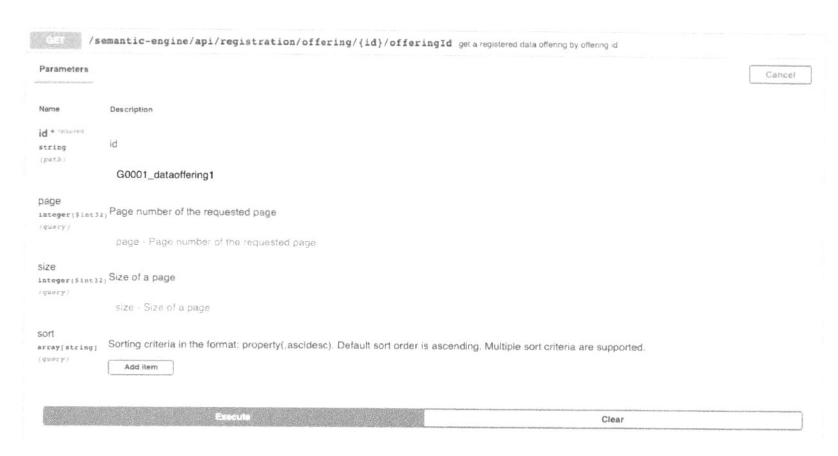

Figure 9.9 Get offering by offering ID.

Query a list of all registered data offerings by provider ID:

The following endpoint is used to fetch all the offerings registered by a data provider – see Figure 9.10. In this endpoint, the user must provide the

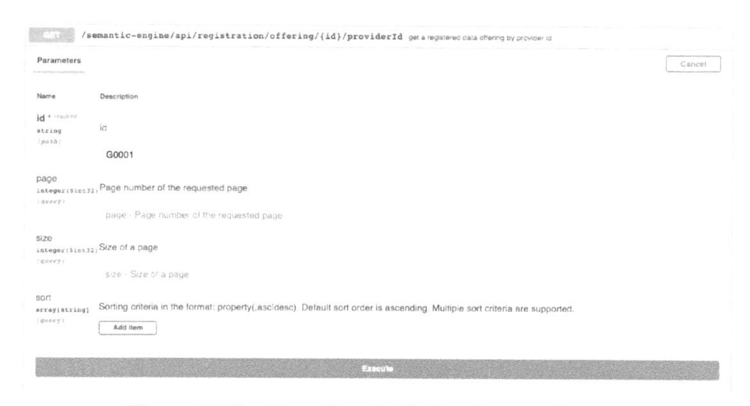

Figure 9.10 Get a list of offerings by provider ID.

"Provider ID". The current release includes retrieval of list of offerings with complete data offering. This might affect the query performance if the data in the storage is increased.

Query a list of all registered data offerings by category:

In the i3-MARKET project, we use different nodes in the cluster and each node has its own semantic engine instance running on it. Furthermore, each instance of semantic engine may have its own data by categories from different pilots (e.g., manufacturing, automotive, wellbeing, etc.) or multiple. Consider a use-case where someone is looking for data offerings registered in i3-MARKET on different nodes. This endpoint allows the user to transparently fetch all the data offerings based on the "category" from i3-MARKET cluster. In summary, this endpoint performs federated query in a distributed nature and brings back the results from different instances in i3-MARKET; see Figure 9.11.

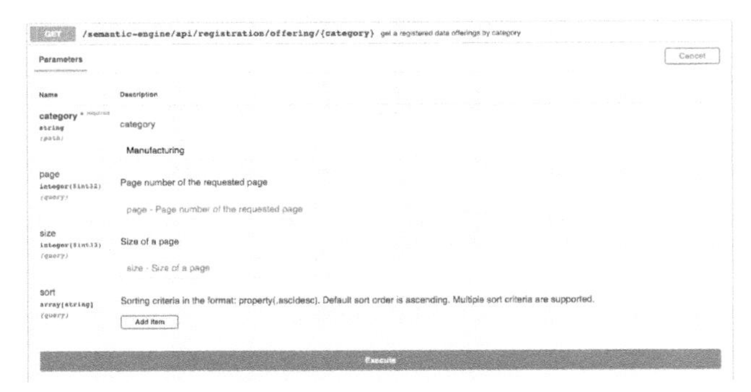

Figure 9.11 Get a list of offerings by category.

Update a data offering:

The following endpoint is used to update an already registered data offering.

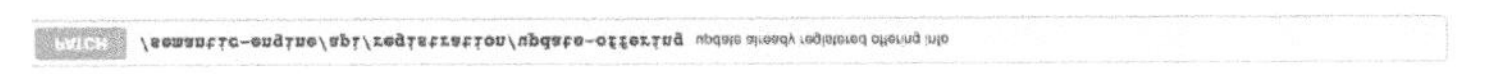

For example, if a specific user can update any field which s/he wants to update, it is important that the user do not change/update the fields with -id attributes, e.g., dataOfferingId, pricingModelId, etc., because such attributes are used internally by the semantic engine to link the data.

Delete a data offering:

This endpoint can be used to permanently remove an offering from the repository. The user must provide the "Offering ID" of the data offerings they want to delete; see Figure 9.12.

Figure 9.12 Delete offering by ID.

Download data offering template:

Figure 9.13 shows that endpoint is used to download the offering template.

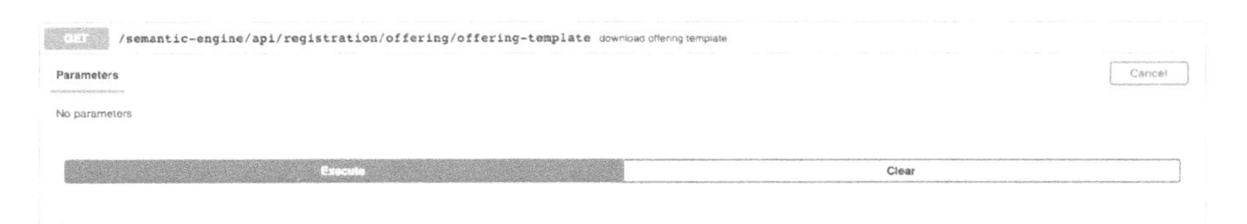

Figure 9.13 Get data offering template.

Query list of offerings by active state:

Figure 9.14 shows an endpoint used to search data offerings that are "active" and so made available to be seen and searched by their providers.

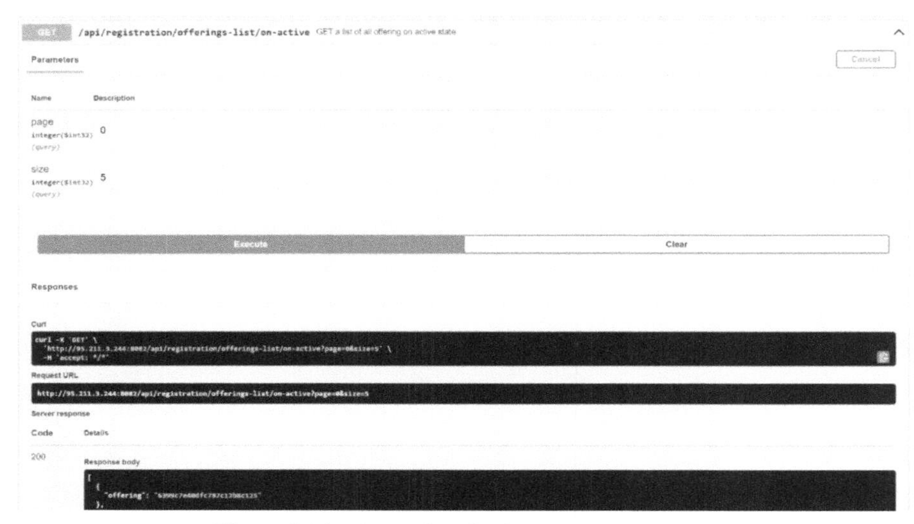

Figure 9.14 Query list of offerings by active state.

Query list of offerings by shared state:

Figure 9.15 shows the endpoint to look for data offerings that are set or not available to be shared in the network by the data marketplaces.

Figure 9.15 Query list of offerings by shared state.

Query offerings based on text/keyword:

Figure 9.16 shows an endpoint can be used for text searches.

Figure 9.16 Query offerings based on text/keyword.

Query offerings in federated network:

The semantic engine is able to search, discover, and retrieve data offerings not only in single instance of a marketplace but also throughout the entire nodes of marketplace belonging to the i3-MARKET network via federated queries. This is possible using the information that each semantic engine manages via SEED-INDEX in the shared BESU blockchains, where there is info about each node/engine. With such details, each semantic engine can search and retrieve the “shared, active” data offerings from the other data marketplace to be consulted by consumers and expanding the availability of offered assets from one marketplace to the entire network.

Following are some of the endpoints:

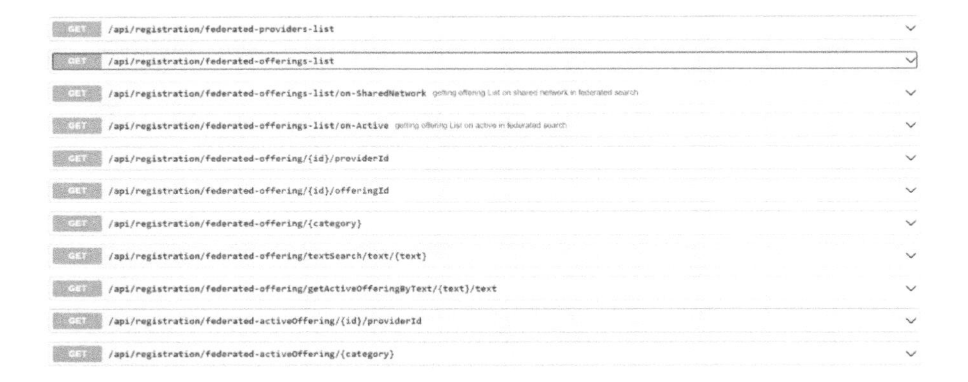

To be noted, most of the endpoints work the same as in local node. The main difference is that now we can search from the cluster or network of registered marketplaces/endpoints.

SDK(-RI) semantic engine services:

Once the functionalities of the semantic engine from the internal API interfaces are mapped and reflected in the Backplane API gateway, they are available to be used via the i3-MARKET development kits in languages like Java and JavaScript (among the others) using directly the SDK-Core and/or the SDK-RI services.

9.7 Background Technologies

Data Catalogue Vocabulary (DCAT) – Version 3:

W3C (World Wide Web Consortium) recommendation:

DCAT is an RDF vocabulary designed to facilitate interoperability between data catalogues published on the Web. This document defines the schema and provides examples for its use.

DCAT enables a publisher to describe datasets and data services in a catalogue using a standard model and vocabulary that facilitates the consumption and aggregation of metadata from multiple catalogues. This can increase the discoverability of datasets and data services. It also makes it possible to have a decentralized approach to publishing data catalogues and makes federated search for datasets across catalogues in multiple sites possible using the same query mechanism and structure.

https://www.w3.org/TR/vocab-dcat-3/

Also, its extension DCAT-AP:

The DCAT Application Profile for data portals in Europe (DCAT-AP) is a specification based on the Data Catalogue Vocabulary (DCAT) developed by W3C.

This application profile is a specification for metadata records to meet the specific application needs of *data portals in Europe* while providing semantic interoperability with other applications on the basis of reuse of established controlled vocabularies (e.g., EuroVoc) and mappings to existing metadata vocabularies (e.g., Dublin Core, SDMX, INSPIRE metadata, etc.).

DCAT-AP provides a common specification for describing public sector datasets in Europe to enable the exchange of descriptions of datasets among data portals. DCAT-AP allows:

- *Data catalogues* to describe their dataset collections using a standardized description, while keeping their own system for documenting and storing them.
- *Content aggregators*, such as the European Data Portal, to aggregate such descriptions into a single point of access.
- *Data consumers* to find datasets more easily through a single point of access.

DCAT-AP has an extension GeoDCAT-AP for describing geospatial datasets, dataset series and services. Another extension, StatDCAT-AP, provides specifications and tools that enhance interoperability between descriptions of statistical datasets within the statistical domain and between statistical data and open data portals.

https://joinup.ec.europa.eu/solution/dcat-application-profile-data-portals-europe/release/200

RDF Store

As part of the marketplace persistence framework back-end layer, we need to use and deploy a database that is able to store our semantic (meta)data in the best way. This database represents the main registry and repository where all the semantically annotated (meta)data are uploaded and saved.

In the persistent database, it is needed to store all the (meta)data descriptions created and collected by marketplace stakeholders, e.g., with the information about providers, consumers, offering descriptions, and recipes. In our research for a semantic interoperability in i3-MARKET, we decided to model our providers, consumers, data offering descriptions, and parameters following an RDF schema model, annotated with our *i3-MARKET Semantic Core Model* and represent and exchange data in JSON serialization format.

So due to the nature of such kind of (meta)data, we need to choose the best solution for storing, managing, accessing, and retrieving information.

RDF triple-store is a type of graph database that stores semantic facts. Being a graph database, triple-store stores data as a network of objects with materialized links between them. This makes RDF triple-store a preferred choice for managing highly interconnected data. Triple-stores are more flexible and less costly than a relational database, for example.

The RDF database, often called a semantic graph database, is also capable of handling powerful semantic queries and of using inference for uncovering new information out of the existing relations. In contrast to other types of graph databases, RDF triple-store engines support the concurrent storage of data, metadata, and schema models (e.g., the so-called ontologies). Models/ontologies allow for the formal description of the data. They specify both object classes and relationship properties, and their hierarchical order as we use our i3-MARKET models to describe our resources.

This allows creating a unified knowledge base grounded in common semantic models that allow combining all metadata coming from different sources, making them semantically interoperable to:

- create coherent queries independently from the source, format, date, time, provider, etc.;
- enable the implementation of more efficient semantic querying features;
- enrich the data and make it more complete, more reliable, and more accessible;
- enable to perform inference as triple materialization from some of the relations.

In the following paragraphs, we are going to give some more information and examples about the semantic data formalization, query interface, and the interface of the semantic framework backend layer within the Backplane.

MongoDB

MongoDB is a source-available cross-platform document-oriented database program. Classified as a NoSQL database program, MongoDB uses

JSON-like documents with optional schemas. MongoDB is developed by MongoDB Inc. (https://www.mongodb.com/).

Main features:

- Ad-hoc queries:

MongoDB supports field, range query, and regular-expression searches. Queries can return specific fields of documents and also include user-defined JavaScript functions. Queries can also be configured to return a random sample of results of a given size.

- Indexing:

Fields in a MongoDB document can be indexed with primary and secondary indices or index.

- Replication:

MongoDB provides high availability with replica sets. A replica set consists of two or more copies of the data. Each replica set member may act in the role of primary or secondary replica at any time. All writes and reads are done on the primary replica by default. Secondary replicas maintain a copy of the data of the primary using built-in replication. When a primary replica fails, the replica set automatically conducts an election process to determine which secondary should become the primary. Secondaries can optionally serve read operations, but that data is only eventually consistent by default.

- Load balancing:

MongoDB scales horizontally using sharding. The user chooses a shard key, which determines how the data in a collection will be distributed. The data is split into ranges (based on the shard key) and distributed across multiple shards. (A shard is a master with one or more replicas.) Alternatively, the shard key can be hashed to map to a shard – enabling an even data distribution.

Semantic data model and serialization formats:

Linked data is based around describing real-world things using the resource description framework (RDF). The following paragraphs introduce the basic data model and then outline existing formats to serialize semantic data models.

The semantic descriptions are generated following the *i3-MARKET Core Model*, annotated with the *i3-MARKET Domain Models*, and mapped with the *i3-MARKET Application Model* vocabularies and then loaded into a registry-store.

Semantic data model:

Figure 9.17 represents an RDF triple. RDF is very simple, flexible, and schema-less to express and process a series of simple assertions. Consider the following example: "Sensor A measures 21C". Each statement, i.e., piece of information, is represented in the form of *triples* (RDF triples) that link a *subject* ("Sensor A"), a *predicate* ("measures"), and an *object* ("21C"). The subject is the thing that is described, i.e., the resource in question. The predicate is a term used to describe or modify some aspect of the subject. It is used to denote relationships between the subject and the object. The object is, in RDF, the "target" or "value" of the triple. It can be another resource or just a literal value such as a number or word.

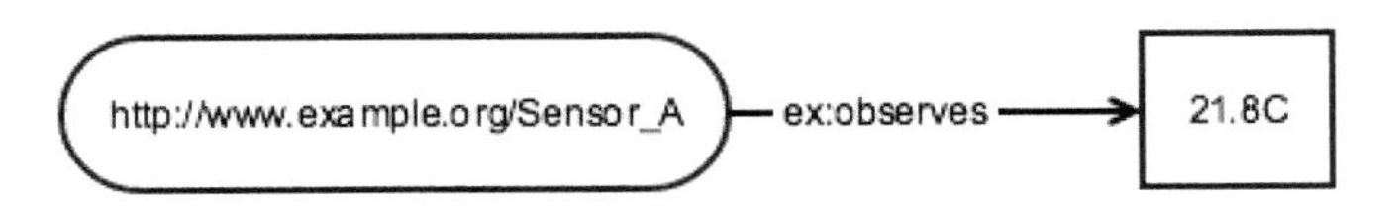

Figure 9.17 RDF triple in graph representation describing "Sensor A measures 21.8°C".

Since objects can also be a resource with predicates and objects on their own, single triples are connected to a so-called RDF graph. In terms of graph theory, the RDF graph is a labelled and directed graph. As the illustration, we extend the previous example, replacing the literal "21.8C" by a resource "measurement" for the object in the RDF triple in Figure 9.18. The resource itself has two predicates assigning a unit and the actual value to the

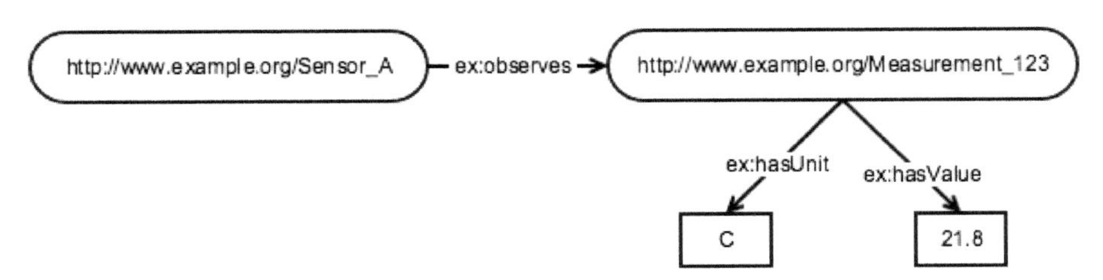

Figure 9.18 Simple RDF graph including the example RDF triple.

measurement – see Figure 9.18. The unit is again represented by a resource and the value is numerical literal. The resulting RDF graph looks as follows:

Serialization formats:

The RDF data model itself does not describe the format in which the data, i.e., the RDF graph structure, is stored, processed, or transferred. Several formats exist that serialize RDF data; the following overview lists the most popular formats, including a short description of their main characteristics and examples. Figure 9.18 shows a simple RDF graph to serve as the basis.

RDF/XML:

The RDF/XML syntax is standardized by the W3C and is widely used to publish linked data on the Web. On the downside, however, the XML syntax is also viewed as difficult for humans to read and write. This recommends consideration of:

a) other serialization formats in data management and control workflows that involve human intervention;
b) the provision of alternative serializations for consumers who may wish to examine the raw RDF data.

The RDF/XML syntax is described in detail as part of the W3C RDF Primer. The MIME type that should be used for RDF/XML within HTTP content negotiation is application/rdf+xml. The listing below shows the RDF/XML serialization for the RDF graph.

RDF/XML serialization example:

```
<?xml version="1.0"?>
<rdf:RDF xmlns:ex="http://www.example.org/"
<rdf:Description rdf:about=" http://www.example.org/Sensor_A">
        <ex:title>21.8°C</ex:title>
</rdf:Description>
</rdf:RDF>
```

Turtle: Turtle (Terse RDF Triple Language) is a plain text format for serializing RDF data. It has support for namespace prefixes and other shorthands, making Turtle typically the serialization format of choice for reading RDF triples or writing them by hand. A detailed introduction to Turtle is given in the W3C Team Submission document Turtle. It was accepted as a first working draft by the World Wide Web Consortium (W3C) RDF Working Group in August 2011, and parsing and serializing RDF data is supported by a large number of RDF toolkits. The following listing shows the serialization listing for the example RDF graph in Turtle syntax.

Turtle serialization example:

```
@prefix : <http://www.example.org/> .
:Sensor_A :measures "21.8°C"
```

N-Triples: The N-Triples syntax is a subset of Turtle, excluding features such as namespace prefixes and shorthands. Since all URIs must be specified in full in each triple, this serialization format involves a lot of redundancy, typically resulting in large N-Triples particularly compared to Turtle, but also to RDF/XML. This redundancy, however, enables N-Triples files to be parsed one line at a time, benefitting the loading and processing of large data files that will not fit into main memory. The redundancy also allows compressing N-Triples files with a high compression ratio, thus reducing network traffic when exchanging files. These two factors make N-Triples the *de facto* standard for exchanging large dumps of linked data. The complete definition of the N-Triples syntax is given as part of the W3C RDF test cases recommendation. The following listing in Table 7.1 represents the N-Triples serialization of the example RDF graph.

N-Triples serialization example:

```
<http://www.example.org/Sensor_A> <http://www.example.org/measures> "21.8°C"@en-UK
.
```

JSON-LD: Used as main data model for the metadata in i3-MARKET.

Jason-LD (https://json-ld.org/) – Many developers have little or no experience with linked data, RDF, or common RDF serialization formats such as N-Triples and Turtle. This produces extra overhead in the form of a steeper learning curve when integrating new systems to consume linked data. To counter this, the project consortium decided to use a format based on a common serialization format such as XML or JSON. Thus, the two remaining options are RDF/XML and JSON-LD. JSON-LD was chosen over

RDF/XML as the data format for all linked data items in BigIoT. JSON-LD is a JSON-based serialization for linked data with the following design goals:

- **Simplicity:** There is no need for extra processors or software libraries, just the knowledge of some basic keywords.
- **Compatibility:** JSON-LD documents are always valid JSON documents; so the standard libraries from JSON can be used.
- **Expressiveness:** Real-world data models can be expressed because the syntax serializes a directed graph.
- **Terseness:** The syntax is readable for humans and developers need little effort to use it.
- **Zero edits:** Most of the time JSON-LD can be devolved easily from JSON-based systems.
- **Usable as RDF:** JSON-LD can be mapped to/from RDF and can be used as RDF without having any knowledge about RDF.

From the above, terseness and simplicity are the main reasons that JSON-LD was chosen over RDF/XML. JSON-LD also allows for referencing external files to provide context. This means contextual information can be requested on demand and makes JSON-LD better suited to situations with high response times or low bandwidth usage requirements. More information can be found in http://json-ld.org/.

The data model underlying JSON-LD is a labelled, directed graph. There are a few important keywords, such as @context, @id, @value, and @type. These keywords are the core part of JSON-LD. Four basic concepts should be considered:

- **Context:** A context in JSON-LD allows using shortcut terms to make the JSON-LD file shorter and easier to read (as well as increasing its resemblance with pure JSON). The context maps terms to IRIs. A context can also be externalized and reused for multiple JSON-LD files by referencing its URI.
- **IRIs:** Internationalized resource identifiers (IRIs) are used to identify nodes and properties in linked data. In JSON-LD, two kinds of IRIs are used: absolute IRIs and relative IRIs. JSON-LD also allows defining a common prefix for relative IRIs using the keyword @vocab.
- **Node identifiers:** Node identifiers (using the keyword @id) reference nodes externally. As a result of using @id, any RDF triples produced for this node would use the given IRI as their subject. If an application follows this IRI, it should be able to find some more information about

the node. If no node identifier is specified, the RDF mapping will use blank nodes.

- **Specifying the type:** It is possible to specify the type of a distinct node with the keyword @type. When mapping to RDF, this creates a new triple with the node as the subject, a property rdf:type and the given type as the object (given as an IRI).

JSON-LD example:

```
[{"@id":"http://www.example.org/Sensor_A","http://www.example.org/measures":[{"@val
ue":"21.8C"}]}]
```

SPARQL:

SPARQL (SPARQL protocol and RDF query language, https://www.w3.org/TR/sparql11-query/) is the most popular query language to retrieve and manipulate data stored in RDF and became an official W3C recommendation in 2008. Depending on the purpose, SPARQL distinguishes the following for query variations:

- **SELECT query:** Extraction of (raw) information from the data.
- **CONSTRUCT query:** Extraction of information and transformation into RDF.
- **ASK query:** Extraction of information resulting a true/false answer.
- **DESCRIBE query:** Extraction of RDF graph that describes the resources found.

Given that RDF forms a directed, labelled graph for representing information, the most basic construct of a SPARQL query is a so-called *basic graph pattern*. Such a pattern is very similar to an RDF triple with the exception that the subject, predicate, or object may be a variable. A basic graph pattern matches a subgraph of the RDF data when RDF terms from that subgraph may be substituted for the variables and the result is RDF graph equivalent to the subgraph. Using the same identifier for variables also allow combining multiple graph patterns. Besides aforementioned graph patterns, the SPARQL 1.1 standard also supports the sorting (ORDER BY), and the limitation of result sets (LIMIT, OFFSET), the elimination of duplicates (DISTINCT), the formulation of conditions over the value of variables (FILTER), and the possibility to declare a constraint as OPTIONAL. The SPARQL 1.1 standard significantly extended the expressiveness of SPARQL. In more detail, the new features include:

- Grouping (GROUP BY) and conditions on groups (HAVING).
- Aggregates (CONT, SUM, MIN, MAX, AVG, etc.).
- Subqueries to embed SPARQL queries directly within other queries.
- Negation to, e.g., check for the absence of data triples.
- Project expression, e.g., to use numerical result values in the SELECT clause within mathematical formulas and assign new variable names to the result.
- Update statements to add, change, or delete statements.
- Variable assignments to bind expressions to variables in a graph pattern.
- New built-in functions and operators, including string functions (e.g., CONCAT, CONTAINS, etc.), string digest functions (e.g., MD5, SHA1, etc.), numeric functions (e.g., ABS, ROUND, etc.), or date/time functions (e.g., NOW, DAY, HOURS, etc.).

As mentioned previously, RDF graph data is represented as triples, i.e., "subject", "predicate", and "object". A very basic SPARQL, which brings back 100 triples from the RDF graph, can be written as follows.

SPARQL example:

```
SELECT * WHERE {?s ?p ?o} LIMIT 100
```

10

Interfaces

10.1 Data Access API

The data access API is the interface via which data consumers gain access to the data offered by a data provider or data space. Since this open interface enables direct interactions among stakeholders of different data spaces/marketplaces, we need not only an open interface specification that can be implemented by all but also a high level of security, as the data exchange might involve sensitive data, e.g., personal data or commercial data.

The endpoints documented below were grouped by modules.

Batch controller:

Get /batch/listDataSourceFiles/{offeringId}
(**batchController.listDataSourceFiles**)
Returns a list of datasets that are available for consumption.

post /batch/{data}/{agreementId}
(**batchController.getBatch**)
Requests data from a provider in the form of batch.

- **Request – First Block**
- **Request – Intermediate Block**
- **Request – Last Block**

Get /batch/listDataSourceFiles/{offeringId}
(**batchController.listDataSourceFiles**)
Returns a list of data sets that are available for consumption.

post /batch/{data}/{agreementId}
(**batchController.getBatch**)
Requests data from a provider in the form of batch.

- **Request – First Block**
- **Request – Intermediate Block**
- **Request – Last Block**

post /batch/pop

Stream controller:

post /newdata/{offeringId}
(**streamController.newData**)
Endpoint for the data soure to send data to the mqtt broker.

Agreement controller:

post /agreement/payMarketFee/{agreementID}
(**agreementController.payMarketFee**)
Endpoint the consumer can use to pay the market fee.

post /agreement/deployRawPaymentTransaction/{agreementId}
(**agreementController.deployRawPaymentTransaction**)
Endpoint for the consumer to deploy a transaction obtained by signing a transaction object that resulted by paying the market fee.

get /agreement/getAgreementId /{exchangeId}
(**agreementController.getAgreementId**)
Endpoint to retrieve the agreementId.

post /agreement/dataExchangeAgreementInfo
(**agreementController.dataExchangeAgreementInfo**)
Endpoint for the provider to post information relevant for the data transfer.

Connector registration controller:

post /regds
(connectorRegistrationController.regds)
Endpoint used by the provider to register the batch or stream data connectors.

Stream auth controller:

post /stream/auth/user

(**streamAuthController.authStreamUser**)
Endpoint to authenticate data transfer stream user.

post /stream/auth/acl
(**streamAuthController.authStreamAcl**)
Endpoint to check if the topic subscribed by the consumer matches a pre-set description standard.

OIDC auth controller:

get /oidc/login/provider
(**oidcAuthController.oidcLoginProvider**)
Endpoint to retrieve a bearer token as provider.

get /oidc/login/consumer
(**oidcAuthController.oidcLoginConsumer**)
Endpoint to retrieve a bearer token as consumer.

get /oidc/cb
(**oidcAuthController.oidcCb**)
Endpoint that will be called after a successful authentication as either a consumer or a provider.

Data transfer report controller:

post /report/nrpCompletnessCheck
(**dataTransferReportController.nrpCompletenessCheck**)
Endpoint to check if the Non-repudiable Protocol was completed for a block of data.

get /report/getListOfVerificationRequests/{agreementId}
(**dataTransferReportController.getListOfVerificationRequests**)
Endpoint to get all the verification requests for an agreement.

get /report/getSubId/{consumerDid}/{offeringId}
(**dataTransferReportController.getSubId**)
Endpoint to get the subscription ID.

get /report/streamingAccountReport/{subId}
(**dataTransferReportController.streamingAccountReport**)
Endpoint for a consumer to get information about a subscription.

get /report/getAccountSummary /{consumerDid}

(dataTransferReportController.getAccountSummary)
Endpoint to get information about the amount of data transferred for a consumer.

10.2 Background Technologies

Loopback:
Loopback is an open-source solution developed by StrongLoop, an IBM company. It is a framework that enables you to create dynamic end-to-end APIs (RESTful and GraphQL). It is for Node.js and developed in TypeScript, a typed superset of JavaScript. Due to its modular connectors, it can (indeed does) support any DB as well as custom data integrations like blockchains.

This is the technology that was selected for the implementation of the data access API.

10.3 Notifications Manager

The notification manager is the service responsible for allowing the creation and emission of notifications both for users and between services; it also integrates the functionality to allow users to subscribe to topics (categories) of the offers, with the objective of receiving notifications when new offers are created that coincide with those subscribed by the user.

On the other hand, it is possible to create notifications for a certain user and store them, access the stored notifications, mark a notification as read, delete it, etc.

10.4 Notifications as a Service

It allows you to send notifications to other services that are not directly connected to our service, and we do not necessarily know who they are. This happens when, for example, a new offer is created, and a request is sent to create a notification to the service and notify the rest of the services/marketplaces that have subscribed to the queue.

This section can be understood from two points of view:

- The first is that of a service that wishes to send a notification to others for a certain event.
- The second is that of a service that expects to receive notifications to perform some action, such as indexing this event within the service.

For the first case, we only need to follow the section "Create Service Notification".

For the second case, we must follow the section "Services and Queues" in order to register our service within the notification manager and need to have an endpoint in our service to receive the notifications.

Services and queues:

In this section, we explain the concepts of service and queue and indicate the methods to work with them.

A service is determined by a name, to identify our service within the system, a generic endpoint where to receive notifications and a list of queues to which it is subscribed.

A queue has a name that indicates what type of event it handles, and it is possible to indicate a specific endpoint where it will send the notifications; this specific endpoint can be null and, in that case, it will use the generic endpoint of the service.

Types of queues:

The following queues have been implemented within the notification manager:

- **offering.new**
- **offering.update**
- **agreement.accepted**
- **agreement.rejected**
- **agreement.update**
- **agreement.pending**
- **agreement.termination**
- **agreement.claim**

Service management:

This section provides methods to perform the following actions:

- list all registered services;
- get the information of a service through its identifier;
- register a service;
- delete a service.

Listing registered services:
GET /services

Get the information of a service through its identifier:
GET /services/{service_id}

Registering a new service:
POST /services

Deleting a service:
DELETE /services/{service_id}

Queue management:

This section indicates the methods to carry out the following actions: * Register a queue * Get the service queues by identifier * Get the information of a specific queue by identifier * Activate-or-deactivate-a-queue * Delete a queue.

Register a queue:
Can also be called subscribe service to queue.
POST /services/{service_id}/queues

Get the service queues by identifier:
GET /services/{service_id}/queues

Obtain the information of a specific queue by identifier:
GET /services/{service_id}/queues/{queue_id}

Activate or deactivate a queue:
PATCH /services/{service_id}/queues/{queue_id}/activate
PATCH /services/{service_id}/queues/{queue_id}/deactivate

Deleting a queue:
DELETE /services/{service_id}/queues/{queue_id}

Create a service notification:
POST /notification/service
The system will search among all the registered services, those that are subscribed to the queue indicated by the receiver_id, then it will create and send a notification to its registered endpoint.

Create a service notification for a single marketplace
POST /notification/service

The system will search among all the registered services, those that are subscribed to the queue indicated by the receiver_id, and then it will create and send a notification to its registered endpoint.

10.5 Notifications to Users

User notifications are messages that are created and stored to be read by the target users. The purpose of these messages is to notify users that an event relevant to them has occurred.

This section indicates the methods to perform the following actions:

- create a user notification.
- access to notifications.
- modification of the notifications.

Create a user notification:
User notifications are created using a POST method.
POST /notification

Access to notifications:

Once a user notification has been created, it can be accessed using one of the following access methods.

Getting all stored notifications:
This method allows access to all notifications stored in the system, including their identifiers and the information contained within the *message* field.
GET /notification

Get all users unread notifications
GET /notification/unread

Get the notifications of a user (by user ID):
GET /notification/user/{user_id}

Get an unread user notification:
GET /notification/user/{user_id}/unread

Get a notification by ID:
GET /notification/{notification_id}

Modification of the notifications:
The following methods are used to make changes to notifications, such as marking them as read/unread and deleting notifications.

Mark notification as read:
PATCH /notification/{notification_id}/read

Mark notification as unread:
PATCH /notification/{notification_id}/unread

Delete notifications:
The following method is used to delete a notification:
DELETE /notification/{notification_id}

10.6 User Subscriptions

A user can subscribe to categories; these categories are the ones to which the offers registered in the semantic engine belong in order to be notified when a new offer related to the category to which the user is subscribed appears.

This indicates the methods to perform the following actions:

- Create a user-subscription.
    - To create a subscription, it is only necessary to have the identifier of the user who wants to subscribe and the category to which he/she wants to subscribe.
- Access the subscriptions.
    - Once a subscription has been created, it is possible to access it by the following methods.
- Modify subscriptions.
    - It is possible to activate/deactivate a subscription and delete it from the system.

11

Conclusions

Data Economy is commonly referring to the diversity in The use of data to provide social benefits and have a direct impact in people's life, from a technological point of view data economy implies technological services to underpin the delivery of data applications that bring value and addressed the diverse demands on selling, buying and trading data assets. The demand and the supply side in the data is increasing exponentially and it is being demonstrated that the value that the data has today is as relevant as any other tangible and intangible assets in the global economy.

In this second i3-MARKET series book, we further discussed why data is the focus of current technological developments towards digital markets and the meaning of data being the next asset to appear evolved in trading markets. At the same time, it focused on introducing the i3-MARKET technology and the proposed solutions.

In this second i3-MARKET series book, we reviewed the basic technological principles and software best practices and standards for implementing and deploying data spaces and data marketplaces. The book provides a definition for data-driven society as: The process to transform data production into data economy for the people using the emerging technologies and scientific advances in data science to underpin the delivery of data economic models and services.

In this book, we discussed the technology assets that are designed and implemented following the i3-MARKET Backplane reference architecture (RA) that uses open data, big data, IoT, and AI design principles to help data spaces and data marketplaces to focus on today's data-driven society as the trend to rapidly transforming the data perception in every aspect of our activities. Moreover, the series of software assets grouped as subsystems and composed by software artefacts are included and explained in full.

Further, we described i3-MARKET Backplane tools and how these can be used for supporting marketplaces and its components. Next, we provided a description of solutions developed in i3-MARKET as an overview of the potential for being the reference open-source solution to improve data economy across different data marketplaces.

References

[1] "https://en.wikipedia.org/wiki/System_context_diagram," [Online].
[2] P. Kruchten, "Architectural Blueprints — The "4+1" View Model of Software Architecture," IEEE Software 12, November 1995, pp. 42-50.
[3] J. R. a. I. J. G. Booch, UML User Guide, Addison-Wesley Longman, 1998.
[4] "https://leanpub.com/arc42inpractice/read," [Online].
[5] i3-MARKET, "i3M-Wallet monorepo," [Online]. Available: https://github.com/i3-Market-V3-Public-Repository/SP3-SCGBSSW-I3mWalletMonorepo.
[6] Consensys, "MetaMask," [Online]. Available: https://metamask.io/.
[7] "Trust Wallet," [Online]. Available: https://trustwallet.com/.
[8] Exodus, "Exodus Bitcoin & Crypto Wallet," [Online]. Available: https://www.exodus.com/.
[9] T. Voegtlin, "Electrum Bitcoin Wallet," [Online]. Available: https://electrum.org/.
[10] Validated ID, "VIDChain," [Online]. Available: https://www.validatedid.com/vidchain.
[11] Evernym, "Connect.Me Wallet," [Online]. Available: https://www.connect.me/.
[12] IdRamp, "IdRamp," [Online]. Available: https://idramp.com/.
[13] trinsic, "Identity Wallets," [Online]. Available: https://trinsic.id/identity-wallets/.
[14] ConsenSys, "uPort," [Online]. Available: https://www.uport.me/.
[15] "Twala," [Online]. Available: https://www.twala.io/.
[16] ConsenSys, "Serto," [Online]. Available: https://www.serto.id/.

[17] "Veramo - A JavaScript Framework for Verifiable Data | Performant and modular APIs for Verifiable Data and SSI," [Online]. Available: https://veramo.io/.

[18] "OpenTimeStamps, a timestamping proof standard," [Online]. Available: https://opentimestamps.org/.

[19] Y. Du, H. Duan, A. Zhou, C. Wang, M. H. Au and Q. Wang, "Enabling Secure and Efficient Decentralized Storage Auditing with Blockchain," IEEE Transactions on Dependable and Secure Computing, 2021.

[20] Y. Du, H. Duan, A. Zhou, C. Wang, M. H. Au and Q. Wang, "Towards Privacy-assured and Lightweight On-chain Auditing of Decentralized Storage," 2020 IEEE 40th International Conference on Distributed, pp. 201-211, 2020.

[21] H. Yu and Z. Yang, "Decentralized and Smart Public Auditing for Cloud Storage," IEEE 9th International Conference on Software, pp. 491-494, 2018.

[22] J. Shu, X. Zou, X. Jia, W. Zhang and R. Xie, "Blockchain-Based Decentralized Public Auditing for Cloud Storage," IEEE Transactions on Cloud Computing, 2021.

[23] K. Liu, H. Desai, L. Kagal and M. Kantarcioglu, "Enforceable Data Sharing Agreements Using Smart Contracts," 27 04 2018. [Online]. Available: https://arxiv.org/abs/1804.10645.

[24] E. J. Scheid, B. B. Rodrigues, L. Z. Granville and B. Stiller, "Enabling Dynamic SLA Compensation Using Blockchain-based Smart Contracts," in IFIP/IEEE Symposium on Integrated Network and Service Management (IM), 2019.

[25] Ocean Protocol Foundation with BigchainDB GmbH and Newton Circus (DEX Pte. Ltd.), "Ocean Protocol: A Decentralized Substrate for AI Data and Services," 2019.

[26] The European Parliament and the Council of the European Union, "General Data Protection Regulation (GDPR). Directive 95/46/EC," 27 04 2016. [Online]. Available: https://gdpr-info.eu/.

[27] K. Jensen and L. M. Kristensen, Coloured Petri nets: modelling and validation of concurrent systems, Springer Science & Business Media, 2009.

[28] Digital Asset Holdings, "Digital Asset Modelling Language (DAML)," [Online]. Available: https://daml.com/.

[29] A. M. Antonopoulos, Mastering Bitcoin: unlocking digital cryptocurrencies, O'Reilly Media, Inc., 2014.

[30] I. Bashir, Mastering blockchain, Packt Publishing Ltd, 2017.

[31] D. Yaga, P. Mell, N. Roby and K. Scarfone, "Blockchain technology overview," arXiv preprint arXiv:1906.11078, 2019.

[32] S. Rouhani and R. Deters, "Security, performance, and applications of smart contracts: A systematic survey," IEEE Access, vol. 7, pp. 50759-50779, 2019.

[33] L. Jing and L. Zhentian, "A survey on security verification of blockchain smart contracts," IEEE Access, vol. 7, pp. 77894-77904, 2019.

[34] G. Wood, "Ethereum: A secure decentralised generalised transaction ledger," Ethereum Project White Paper, vol. 151, no. 2014, pp. 1-32, 2014.

[35] H. Chen, M. Pendleton, L. Njilla and S. Xu, "A survey on ethereum systems security: Vulnerabilities, attacks, and defenses," ACM Computing Surveys (CSUR), vol. 53, no. 3, pp. 1-43, 2020.

[36] "Hyperledger Besu," [Online]. Available: https://github.com/hyperledger/besu.

[37] "Solidity," [Online]. Available: https://solidity-es.readthedocs.io/.

[38] "BIP-39," 2021. [Online]. Available: https://github.com/bitcoin/bips/blob/master/bip-0039.mediawiki.

[39] i3-MARKET, "i3M-Wallet OpenApi Specification," 2022. [Online]. Available: https://github.com/i3-Market-V3-Public-Repository/SP3-SCGBSSW-I3mWalletMonorepo/blob/public/packages/wallet-desktop-openapi/openapi.json.

[40] W3C, "Decentralized Identifiers (DIDs) v1.0. Core architecture, data model, and representations," W3C Recommendation, 19 07 2022. [Online]. Available: https://www.w3.org/TR/did-core/.

[41] W3C, "Verifiable Credentials Data Model v1.1.," W3C Recommendation, 03 03 2022. [Online]. Available: https://www.w3.org/TR/vc-data-model/.

[42] F. Román García and J. Hernández Serrano, "i3M-Wallet Base Wallet," [Online]. Available: https://github.com/i3-Market-V3-Public-Repository/SP3-SCGBSSW-I3mWalletMonorepo/tree/public/packages/base-wallet.

[43] F. Román García and J. Hernández Serrano, "SW Wallet," [Online]. Available: https://github.com/i3-Market-V3-Public-Repository/SP3-SCGBSSW-I3mWalletMonorepo/tree/public/packages/sw-wallet.

[44] F. Román García and J. Hernández Serrano, "BOK Wallet," [Online]. Available: https://github.com/i3-Market-V3-Public-Repository/SP3-SCGBSSW-I3mWalletMonorepo/tree/public/packages/bok-wallet.

[45] F. Román García and J. Hernández Serrano, "Wallet Desktop," [Online]. Available: https://github.com/i3-Market-V3-Public-Repository/SP3-SCGBSSW-I3mWalletMonorepo/tree/public/packages/wallet-desktop.
[46] J. Hernández Serrano and F. Román García, "Server Walllet," [Online]. Available: https://github.com/i3-Market-V3-Public-Repository/SP3-SCGBSSW-I3mWalletMonorepo/tree/public/packages/server-wallet.
[47] J. Hernández Serrano and F. Román García, "Wallet Desktop OpenAPI," [Online]. Available: https://github.com/i3-Market-V3-Public-Repository/SP3-SCGBSSW-I3mWalletMonorepo/tree/public/packages/wallet-desktop-openapi.
[48] F. Román García and J. Hernández Serrano, "Wallet Protocol," [Online]. Available:https://github.com/i3-Market-V3-Public-Repository/SP3-SCGBSSW-I3mWalletMonorepo/tree/public/packages/wallet-protocol.
[49] F. Román García and J. Hernández Serrano, "Wallet Protocol API," [Online]. Available:https://github.com/i3-Market-V3-Public-Repository/SP3-SCGBSSW-I3mWalletMonorepo/tree/public/packages/wallet-protocol-api.
[50] F. Román García and J. Hernández Serrano, "Wallet Protocol Utils," [Online]. Available:https://github.com/i3-Market-V3-Public-Repository/SP3-SCGBSSW-I3mWalletMonorepo/tree/public/packages/wallet-protocol-utils.
[51] IDEMIA, "Video proving the integration of IDEMIA's HW Wallet into the i3-MARKET Wallet Desktop application," 2022. [Online]. Available: https://drive.google.com/file/d/1Ai_eoDIzIHczOjzOMBR4ctV5kbR05NOE/view?usp=share_link.
[52] Bluetooth SIG - Core Specification Workgroup, "Bluetooth Core Specification v2.1 + EDR: Secure Simple Pairing," 2007.
[53] D. Basin, C. Cremers, J. Dreier, S. Meier, R. Sasse and B. Schmidt, "Tamarin Prover," [Online]. Available: http://tamarin-prover.github.io/.
[54] OpenJS Foundation, "Electron," [Online]. Available: https://www.electronjs.org/.
[55] Ethers JS, "The Ethers Project," [Online]. Available: https://github.com/ethers-io/ethers.js/.
[56] Veramo, "Veramo - A JavaScript Framework for Verifiable Data," [Online]. Available: https://veramo.io/.
[57] OpenAPI, "OpenAPI Initiative," Linux Foundation, [Online]. Available: https://www.openapis.org/.
[58] "Express OpenAPI Validator," [Online]. Available: https://github.com/cdimascio/express-openapi-validator.

[59] TypeDoc, "TypeDoc," [Online]. Available: https://typedoc.org.

[60] J. Hernández Serrano, "i3-MARKET Non-Repudiation Library," 2022. [Online]. Available: https://github.com/i3-Market-V3-Public-Repository/SP3-SCGBSSW-CR-NonRepudiationLibrary.

[61] J. Hernández Serrano, "i3-MARKET Conflict Resolver Service," 2022. [Online]. Available: https://github.com/i3-Market-V3-Public-Repository/SP3-SCGBSSW-CR-ConflictResolverService.

[62] J. Hernández Serrano, "API of the i3-MARKET Non-Repudiation Library," i3-MARKET, 2022. [Online]. Available:https://github.com/i3-Market-V3-Public-Repository/SP3-SCGBSSW-CR-NonRepudiationLibrary/blob/public/docs/API.md.

[63] Panva, "JOSE," [Online]. Available: https://github.com/panva/jose.

[64] Ajv, "Ajv JSON schema validator," [Online]. Available: https://ajv.js.org/.

[65] OpenJS Foundation, "Express JS," [Online]. Available: https://expressjs.com/.

[66] Y. Kovacs, S. Stanhke and J. L. Muñoz, "i3-MARKET Smart Contracts," [Online]. Available: https://github.com/i3-Market-V3-Public-Repository/SP3-SCGBSSW-I3mSmartContracts.

[67] Hans van der Veer and Anthony Wiles, "Achieving Technical Interoperability - the ETSI Approach," in ETSI, 2008.

[68] Mike Ushold, Christopher Menzel, and Natasha Noy. Semantic Integration & Interoperability Using RDF and OWL. [Online]. https://www.w3.org/2001/sw/BestPractices/OEP/SemInt/.

[69] M. Compton et al., "The SSN ontology of the W3C semantic sensor network incubator group," JWS, 2012.

[70] EUROPA. Publications Office of the EU. EU Vocabularies. Controlled Vocabularies. Authority tables. Frequency. https://publications.europa.eu/en/web/eu-vocabularies/at-dataset/-/resource/dataset/frequency.

[71] EUROPA. Publications Office of the EU. EU Vocabularies. Controlled Vocabularies. Authority tables. File type. https://publications.europa.eu/en/web/eu-vocabularies/at-dataset/-/resource/dataset/file-type.

[72] EUROPA. Publications Office of the EU. EU Vocabularies. Controlled Vocabularies. Authority tables. Language. https://publications.europa.eu/en/web/eu-vocabularies/at-dataset/-/resource/dataset/language/.

[73] EUROPA. Publications Office of the EU. EU Vocabularies. Controlled Vocabularies. Authority tables. Corporate body. https://publications.europa.eu/en/web/eu-vocabularies/at-dataset/-/resource/dataset/corporate-body/.

[74] EUROPA. Publications Office of the EU. EU Vocabularies. Controlled Vocabularies. Authority tables. Continent https://publications.europa.eu/en/web/eu-vocabularies/at-dataset/-/resource/dataset/continent.

[75] EUROPA. Publications Office of the EU. EU Vocabularies. Controlled Vocabularies. Authority tables. Country. https://publications.europa.eu/en/web/eu-vocabularies/at-dataset/-/resource/dataset/country.

[76] EUROPA. Publications Office of the EU. EU Vocabularies. Controlled Vocabularies. Authority tables. Place. https://publications.europa.eu/en/web/eu-vocabularies/at-dataset/-/resource/dataset/place.

[77] European Commission. Joinup. Asset Description Metadata Schema (ADMS). https://joinup.ec.europa.eu/solution/asset-description-metadata-schema-adms.

[78] CI/CD with Ansible Tower and GitHub. Available from: https://keithtenzer.com/2019/06/24/ci-cd-with-ansible-tower-and-github/.

[79] Red Hat Ansible Tower Monitoring: Using Prometheus + Node Exporter + Grafana. Available from: https://www.ansible.com/blog/red-hat-ansible-tower-monitoring-using-prometheus-node-exporter-grafana.

Index

About the Editors

Dr. Martín Serrano is a recognized expert on semantic interoperability for distributed systems due to his scientific contribution(s) to using liked data and semantic formalisms like ontology web language for the Internet of Things and thus store the collected sensor's data in the Cloud. He has also contributed to define the data interplay in edge computing using the linked data paradigm; in those works he has received awards recognizing his scientific contributions and publications. Dr. Serrano has advanced the state of the art on pervasive computing using semantic data modelling and context awareness methods to extend the "autonomics" paradigm for networking systems. He has also contributed to the area of information and knowledge engineering using semantic annotation and ontologies for describing data and services relations in the computing continuum. Dr. Serrano has defined the data continuum and published several articles on data science and Internet of Things science and he is a pioneer and visionary on proposing that semantic technologies applied to policy-based management systems can be used as an approach to produce cognitive applications capable of understanding, service and application events, controlling the pervasive services life cycle. A process called bringing semantics into the box, as published in one of his academic books. He has published 5 academic books and more than 100 peer reviewed articles in IEEE, ACM and Springer conferences and journals.

Dr. Achille Zappa is a Post-Doctoral Researcher at Insight, University of Galway. He received BSC/MSC degree in Biomedical Engineering and PHD in Bioengineering from the University of Genoa (Italy), his Ph.D. project was related to semantic web integration, knowledge engineering and data management of biomedical and genomic data and his research interests include semantic web technologies, semantic data mashup, linked data, big data management, knowledge engineering, big data integration, semantic integration in life sciences and health care, workflow management, IoT semantic interoperability, IoT semantic data and systems integration. Dr. Zappa is the W3C Advisory Committee representative for Insight Centre at University of Galway and member of W3C working groups like the HCLS IG, the Web of Things (WoT) IG and WG, the Spatial Data on the Web WG. He currently work with the main Insight Linked Data and Semantic Web Groups and with the UIoT (Internet of Things, stream processing and intelligent systems unit) Research Unit, addressing collaboration with different units and involvement in various projects where he seeks to develop general-purpose linked data analytics platform(s), which enables (a) flexible and scalable data integration mechanisms and (b) flexible use and reuse of data analytics components such as visualization components and analytics methods. Dr. Zappa has an extensive expertise of applying semantic web technologies and linked data principles in health care and life sciences domains.

Mr. Waheed Ashraf is a Senior Software Engineer with extensive experience in Java programming with Spring Boot and Project Management experience with a strong background on microservices systems design and is an AWS Certified person. Mr. Ashraf is a highly skilled senior software engineer, with 10+ years of project related professional experience in developing and implementing software systems and developing and maintaining enterprise applications working for international companies from USA, Australia and Malaysia. Mr. Ashraf is also proficient in agile software development, scrum and continuous integration (Jenkins), Amazon Web Services (AWS) and back-end RDBMS (using SQL in Databases Like Oracle, DB2, MySQL 4.0 and Microsoft SQL Server). He is currently responsible for the design, development and implementation of a federated authentication and authorization infrastructure (AAI) for federated access to data providers in the context of the Federated Decentralized Trusted Data Marketplace for Embedded Finance FAME Horizon Europe project.

Dr. Pedro Maló is professor at the Electrotechnical Engineering and Computers Department (DEEC) of the NOVA School of Science and Technology (FCT NOVA), Senior Researcher at UNINOVA research institute and Entrepreneur at UNPARALLEL Innovation research-driven hi-tech SME. He obtained an M.Sc. in Computer Science and holds a Ph.D. in Computer Engineering with research interests in interoperability and integrability of (complex) systems with special emphasis on cyber-physical systems/Internet of Things. Pedro coined novel methods and tools such as the plug'n'play interoperability (PnI) solution for large-scale data interoperability and the NOVAAS (NOVA Asset Administration Shell) that establishes the guidelines and methodology for industry digitization by integrating industrial assets into a Industry 4.0 communication backbone. As an entrepreneur, Pedro initiated the development of the IoT Catalogue that aims to be the whole-earth catalogue of the Internet of Things (IoT) – the one-stop-source for innovations, products, applications, solutions, etc. to help users (developers/integrators/advisors/end-users) to take the most advantage of the IoT for the benefit of society, businesses and individuals. Pedro has 20+ years practice in the management, research and technical coordination/development of RTD and innovation projects in ICT domains especially addressing data technologies, systems' interoperability and integration solutions. Pedro is a recognized Project Manager and S&T Coordinator of European/National RTD and industry projects with skills in the coordination of both co-localized and geographical dispersed work teams operating in multidisciplinary and multicultural environments.

Márcio Mateus is project Manager at Unparallel Innovation, Lda Portugal and a Research engineer holding an M.Sc. in electrotechnical and computer engineering from the Faculty of Science and Technology of the Universidade Nova de Lisboa (FCT NOVA). Márcio is an expert in data interoperability measurement techniques and methodologies for complex heterogeneous environments.

Mr. Edgar Fries is Senior System Architect at Siemens AG, Germany. In his early career he acted as project manager and consultant at SIEMENS AG consulting in the field of engineering with a focus on engineering tools and methods for customers in the plant engineering and product business. Fries is graduated from the Technical computer science in Esslingen University of Applied Sciences.

Iván Martínez is project manager and SW architect at Atos, Spain, and a senior researcher at the ARI department of the company AtoS. He graduated in computer science from Technical University of Madrid and in the past few years he has participated in semantic web, cloud, big data and blockchain related industrial and research projects. He has contributed to national research projects such as PLATA, and other Cloud, HPC and big data related projects, such as KHRESMOI, VELaSCCo, TOREADOR, DataBench and BODYPASS mainly leading in the latter's definition and integration of system architecture.

Mr. Alessandro Amicone is an experience project manager at GFT, Italy leading both public funded and commercial market projects. In the first part of his professional career, he worked mainly in projects focusing on coordinating documents management and business process management systems for the bank and insurance industry. In recent years he has been working on Horizon2020 projects and innovative market projects promoting smart communities and technology for digital transformation for and between companies in the industry sector and research communities. The development of processes and management systems mainly focuses on advancing the state of art using software engineering for blockchain, smart contracts and distributed/self-sovereign identity, ensuring cyber-security solutions.

Justina Bieliauskaite is Innovations Director at the European Digital SME Alliance with more than 8 years of project lead and management experience (previously she worked in Lithuanian and Belgian NGOs). Justina Bieliauskaite leads the preparation and implementation of Horizon Europe, Digital Europe Programme, Erasmus+ and other tenders/service contracts for the European Commission. She is experienced in coordinating stakeholder engagement, policy analysis and recommendations, SME training, standardization, and communication activities. Justina is currently the main coordinator of the BlockStand.eu project. Currently, Justina is leading DIGITAL SME's Projects and Standardisation teams, and coordinates the internal WG DIGITALIZATION which covers AI, IoT, cloud computing, blockchain and emerging technologies, as well as coordination among digital innovation hubs. Justina holds a Master's degree in Science (cum laude), focusing on political science and international relations, from the Universities of Leiden and Vilnius. Besides her mother-tongue Lithuanian, Justina speaks English, Italian, Russian and German.

Dr. Marina Cugurra is a lawyer specializing in R&I projects, in particular in legal issues of new technologies and Information Society (e.g. AI, GDPR, data ownership, etc.), with a Ph.D. degree at the "Telematics and Information Society" Ph.D. School at University of Florence. She is also an expert in ethical and societal themes related to ICT research and technological developments. She is serving as independent Ethical Expert at European Commission and European Defense Agency. Consolidated experience in national projects and international and European projects. Scientific collaboration with CNIT (National Inter-University Consortium for Telecommunications) and CNR – ITTIG (Italian National Research Council, Institute of Legal Information Theory and Techniques). Legal Advisor in the R&I Division of multinational companies. She has contributed to the activities of the legal working groups of Eu-wide initiatives (EU Blockchain Observatory Forum) and is Chair of the Ethics, Data Protection and Privacy (EDPP) Task Force of the "Citizen's Control of Personal Data" Initiative within Smart City Marketplace.